For Koen

The Digital Renaissance of Work

'For once you have tasted flight, you will walk the earth with your eyes turned skywards, for there you have been and there you will long to return.'

Leonardo da Vinci
Artist

The Digital Renaissance of Work

Delivering Digital Workplaces Fit for the Future

Paul Miller
and
Elizabeth Marsh

Routledge
Taylor & Francis Group

LONDON AND NEW YORK

First published 2014 by Gower Publishing

Published 2016 by Routledge
2 Park Square, Milton Park, Abingdon, Oxon OX14 4RN
711 Third Avenue, New York, NY 10017, USA

Routledge is an imprint of the Taylor & Francis Group, an informa business

British Library Cataloguing in Publication Data
A catalogue record for this book is available from the British Library

ISBN: 978-1-4724-3720-4 (pbk)

Library of Congress Cataloging-in-Publication Data
Miller, Paul, 1957-
  The digital renaissance of work : delivering digital workplaces fit for the future / by Paul Miller and Elizabeth Marsh.
    pages cm
  Includes bibliographical references and index.
    ISBN 978-1-4724-3720-4 (hardback) -- ISBN 978-1-4724-3721-1 (ebook) -- ISBN (invalid) 978-1-4724-3722-8 (epub) 1. Technological innovations--Management. 2. Employees--Effect of technological innovations on. 3. Telecommuting. 4. Virtual reality in management. 5. Virtual work teams. I. Title.
  HD45.M5155 2015
  658.3'120285--dc23

                                                                                2014020510

Printed and bound in Great Britain by
TJ International Ltd, Padstow, Cornwall

Contents

List of Figures   *vii*
List of Tables   *ix*
Acknowledgements   *xi*
Acronyms and Abbreviations   *xiii*
Foreword by Brian Solis   *xiv*
Preface by Paul Miller   *xx*
Preface by Elizabeth Marsh   *xxiii*

PART I   EXPLORING THE DIGITAL RENAISSANCE OF WORK

1   **Plentiful Work We Enjoy – A First in Human History**   3

2   **The Human-Centred Digital Workplace**   13

3   **Where Will We Work in the Physical World?**   23

4   **Collaboration – and Why 'Teamwork' Needs a Makeover**   33

5   **There Are No Jobs – But There Is Lots of Work**   39

6   **Leaders Need Followers**   49

7   **The Price We Pay**   55

8   **Education – The Revolution Starts Here**   63

9   **A Future Fit for Work**   73

PART II   DELIVERING DIGITAL WORKPLACES FIT FOR THE FUTURE

10   **Your Digital Workplace Journey**   77

11   **Making the Business Case**   97

12   **Designing for a Flexible Workforce**   **117**

13   **Setting Up the Digital Workplace Programme**   **141**

14   **Measuring Progress and Performance**   **167**

Appendix 1   *Rolling Out Spark, PwC's Transformative Global Collaboration Platform*   *183*
Appendix 2   *The Journey Towards the Cisco Connected Workplace*   *191*
Appendix 3   *Creating a Better Place to Work: Microsoft's Workplace Advantage*
                Programme   *197*
Appendix 4   *Engaging Employees and Joining Up the Organization: The Virgin Media*
                Digital Workplace   *203*

Index   *207*

List of Figures

Figure 13.1   Functions represented as key stakeholders at a strategic or
                ownership level   145
Figure 13.2   Functions represented as members of the project delivery team   147
Figure 13.3   Successful outcomes from working across different functions   148

List of Tables

Table 10.1   The five levels of the digital workplace model, illustrated by
              typical employee and leader reactions   85
Table 10.2   The five levels for 'communication and information'   86
Table 10.3   The five levels for 'community and collaboration'   88
Table 10.4   The five levels for 'services and workflow'   89
Table 10.5   The five levels for 'structure and coherence'   91
Table 10.6   The five levels for 'mobility and flexibility'   92

Table 14.1   Example KPIs and metrics linked to benefits emanating from
              digital workplace initiatives   170
Table 14.2   Examples of KPIs and metrics for individual digital workplace
              projects   175

Acknowledgements

Special Thanks

Special thanks go to our superb editor Alison Chapman, for encouraging grand ideas while at the same time remaining focused on the details, and to our hugely knowledgeable researcher Steve Bynghall for providing a fascinating range of case studies. Also to Ephraim Freed for understanding how to raise awareness across our industry; Nancy Goebel for great help with strategic partnerships; and to Jonathan Norman from Gower, who planted the original concept of 'another digital workplace book' and then guided us through the writing. And there is also a special thank you to Brian Solis for contributing a compelling Foreword that shares the excitement we feel about the Digital Renaissance.

More Thanks

We have been supported generally by colleagues in Digital Workplace Group (DWG)'s research programme, which is acknowledged in the industry for its thought leadership. Over ten years this programme has built up a library of more than 50 cutting-edge reports into intranet and digital workplace practice. Behind this incredible resource is a team of exceptionally knowledgeable researchers and practitioners, whose high-quality work has helped to underpin this book. In particular, we would like to thank Julie Lakha, Chris Tubb and Steve Bynghall, who are all pioneering the digital workplace through their ground-breaking research.

In addition, our thanks go to members of the DWG team who have contributed in various and valuable ways to the production of the book as a whole: Helen Day, Mark Silverman, Louise Bloom, Louise Kennedy, Andrew Marr, Angela Pohl, John Baptista and Alison Newman. Excellent work was also provided by our designer Sean House.

Other grateful acknowledgements go to a huge list of DWG members and clients, industry practitioners and thought leaders, too many to mention individually. But particular thanks are due to the digital workplace professionals who have contributed

ROI	Return On Investment
SaaS	Software-as-a-Service
SME	Small and Medium-size Enterprise
SMS	Short Message Service
SMT	Sustainable Management Team
SOLE	Self-Organized Learning Environment
USPTO	US Patent and Trademark Office
VCoIP	Video Conferencing over Internet Protocol
VPN	Virtual Private Network

Foreword

Brian Solis

The Future of Work Begins with a Renaissance in Digital Philosophy

The future of work is an increasingly popular discussion topic among technologists, executives and those managing human capital. When attempting to peer into the future, entirely new worlds of possibilities emerge. That sounds promising but it presents decision-makers with a dilemma. Each vision of the future is influenced by our own interpretation of current trends. How we perceive these trends is, of course, based on our individual experiences and intuition, as well as our hopes and aspirations for what might happen in the future.

But for all the talk about the future, most visions are unfortunately anchored in the past. And without understanding this disconnect between the past and the future, we cannot build a strong bridge to what lies ahead. I believe that what lies ahead is a revolution, not only in how we work but in the ways we think about work. This is why Paul Miller and Elizabeth Marsh's *Digital Renaissance of Work* is so essential at this moment. In exploring both the digital and physical aspects of work, the authors put the emphasis on the human beings at the center of this digital transformation.

Change either happens to us or because of us. Together we can help usher in a future that will work *for* us – rather than simply allowing the future to be forced upon us. Let me be clear … change starts with *us*.

> *Everyone thinks of changing the world, but no one thinks of changing himself.*
> Leo Tolstoy

The Digital Renaissance is a Revival of Basic Human Principles

The mistake most futurists make is predicting the future in terms of technology rather than looking at the effect of technology on human behavior. Without understanding how societal factors impact our preferences, expectations and values, there is a disconnection between aspiration and realization.

Technology is connecting us in ways that improve communication, discovery and connectivity. The world is becoming much smaller as a result. Chances are that you are connected in one network or another, to people in at least 12 other countries. Although social networking and smartphones are relatively new in the everyday life of adults and kids, the way we use these networks and devices as consumers is outpacing the way we use technology in the workplace as employees.

Over time, how we make decisions as consumers, what we come to expect from the companies we do business with and how we want to work with companies are bringing about an end to 'business as usual'; pushing those who cherish normalcy outside their comfort zones. In this 'post-normal' world, power is shifting away from today's business practices and the models that support the status quo towards free-thinking, democratized markets, in which information and people are equally connected.

Unfortunately, companies are weighed down by years, or even decades, of technology investment – and by the existing philosophies and processes that govern and support those investments. But what are the implications if technology and society continue to evolve faster than the ability of businesses to adapt? This scenario sets the stage for what I call 'Digital Darwinism', where businesses become extinct due to the pace of digital and societal change.

But it doesn't stop there. Today's connected customers, driving change in connected markets, aren't the only ones we need to understand. They also represent a growing percentage of our workforce. And with this rise of connected employees, a separation appears between the connected and the less connected, and between the principles, expectations and aspirations associated with each faction.

The Great Divide: Intellectual Property is Ageing While New IP is Under-Capitalized

> *There is a widening skills gap where the existing workforce has been educated and trained to obtain the jobs of yesterday and not the jobs of today and tomorrow.*
> *Jeff Weiner, CEO at LinkedIn*[1]

As a result of technology's impact on society, people have become more informed. Information often leads to empowerment. And empowerment leads people to become more aware and more demanding. In addition, connectedness has created an 'always-on' society that lives in real-time, where the line between the real and virtual worlds blurs to the point of oneness.

Currently, there is a striking gap – and it's widening – in how we manage employees versus how they work or want to work. Organizations are standing at a critical inflection point where one route continues along a comfortable path of business as usual, while the other is unpaved, and without waypoints. It requires those who traverse it to chart the way ahead for themselves.

As this compelling book says, the Digital Renaissance of Work requires a new understanding of how you regard the value of your human capital.

Everyone has something to learn.

When we think we know enough, we stop learning. Yet, the world continues to evolve with or without us.

Fighting Fire with Fire Will Only Burn Everything to the Ground

In my research, I've found that many executives are well aware of the onslaught of new technology. Most already feel the pain of trying to manage the differences between ageing and younger workforces while leading both towards productivity and success. At the same time, many are unsure of how to solve the problem or even assess what the problem really is. How is the future of work different to the present situation?

There are those in IT who are drafting new plans that alter long-established roadmaps to evaluate emergent social and mobile technologies. Some are bolting trendy technologies onto legacy systems in order to apply what will prove to be only a temporary fix. As my friend Stowe Boyd, a web anthropologist and futurist, often says: 'You can't teach old tech new tricks.'

Either way, social and mobile threw a curveball. It wasn't just because the technology overtook the world in a matter of a few short years; it's that social media and mobile apps have changed the behavior of people who use them. Suddenly, businesses have to rethink … everything. Yet, their structure still symbolizes the old guard of the 'command and control' approach, where employees used technology bestowed upon them because it was gospel. In today's world though, all I can say is, 'Good luck with that strategy.' More often than not, the technology we force onto people pushes them to conform to a way of work dictated by the technology rather than to use the technology as a seamless enabler to get work done, individually or collectively – in the way people use technology naturally in their personal lives.

Throwing technology at the problem isn't the answer to enlivening the future or embracing this renaissance.

Technology is Part of the Solution but it's Also Part of the Problem

As a digital analyst and anthropologist, I explore the dynamics of human behavior from a bottom-up or escalation perspective. The conundrum facing IT and businesses overall, is that the philosophies and systems governing the way we work are traditionally designed from a top-down approach.

As technology was introduced into our work, it was usually from a centralized process, typically to control purchase and management costs, and to contain rogue installations within departments. With the best intentions, I'm sure, technology was meant to increase efficiencies, reduce long-term costs and improve the dynamics for working, collaborating and learning.

Meanwhile, as cost and adoption barriers fell, technology worked its way into consumer homes. The result is that consumer familiarity with and receptiveness to technology have outpaced those of companies and institutions.

People are learning, communicating and collaborating differently in their personal lives. And yet, elsewhere, they are still expected to follow dated protocols that are at best counterintuitive. Connected employees are fueling an escalation of expectations and demands to do things differently, while decision-makers struggle to figure out why investments don't pan out according to plan. Older workforces are left wondering why younger counterparts just can't follow the rules, tending always to compare things to the way the world was rather than look to the world that is evolving, with emergent technology viewed as a mere novelty or just frustrating.

It's almost as if protestors expect the world to regress rather than progress.

The only way to take a meaningful step forward is to understand how to adapt legacy investments, systems and processes to pave the way for a more engaging and productive future for all.

Managers Manage Employees Against the Past, Not the Future

Every year, Gallup publishes a comprehensive study that measures the state of employee happiness. In 2013, the study revealed abysmal results: employee morale, Gallup found, is at an 'all-time low'. Only 13 per cent of today's workers feel engaged by their jobs.[2] An astounding 24 per cent though are 'actively disengaged', meaning they show up to work to get a paycheck but couldn't care less about the company or their managers and co-workers. As if that wasn't bad enough, around 50 per cent are 'disengaged', because they are not 'inspired by their managers'.

In his blog, Jim Clifton, Gallup's chairman and CEO, wrote that poor management is one of the leading causes of employee disengagement.[3] Managers are managing against a set of rules and processes that symbolize the way the world used to work. They do not mirror the freedom and empowerment that has been unlocked because of the democracy of social technology.

Most interestingly, Gallup found that those allowed to work remotely were more engaged – a finding that goes contrary to Yahoo! CEO Marissa Meyer's notorious ordinance to stop all telecommuting of employees.[4]

Technology is Most Effective When it is Invisible

No matter how innovative the technology, it won't count for much if we don't define a vision of what we are trying to do – something that will matter to people. Without this, we are just operating under business as usual … the way we always have. What Paul Miller and Elizabeth Marsh do is powerfully describe a vision of a human-centered digital world of work. This is vital as, currently, we are often taking something new and trying to force it into a system that in effect strips away the very essence of what makes it revolutionary. Thus, we cannot move in any new or meaningful directions until we change our philosophy and governance around technology and what the future of work actually means, what it offers, and what we can actually do with it … together.

What is key to understand is how people are using technology and how their behaviors, values and expectations have evolved. In this way, technology begins to look like an enabler for something more natural.

It is crucial to view people not through a lens of scepticism but instead through one of possibility. Then you'll find that the very essence of social and mobile is tied less to tech and more to the lack of hierarchies and absence of 'command and control' dictatorships. The tools become democratized, designed 'by the people for the people'.

They can be properly SOCIAL, equalized for those who use them. They are flat in how people connect and share. This is the real driver for social and mobile today. Systems architects and managers need to understand behavior and to empathize with it in order to be inspired to lead it in meaningful ways.

We Need Leaders, Not Managers

As this book stresses in Chapter 6 on leadership, this is a powerful time for leaders … but not for the conventional management systems as we know them. Change doesn't have only to come from today's executives or managers, however. What is important to understand is that change can come from anywhere within the organization. Anyone can assume the role of leader as long as they have a vision for what is possible, the courage to break what isn't yet fully broken and the passion to mobilize people to unite in transformation. This sense of conviction is contagious and, when approached with a human and business focus, even executives cannot help but listen … and learn. I guess that's what this is all about. We have to learn to learn again and that will help us to lead.

The Digital Renaissance begins with you … and is because of you.

Brian Solis, Principal, Altimeter Group, Digital Analyst and Anthropologist, Futurist,
Author, Human
@briansolis
www.briansolis.com

Notes

1   Weiner, Jeff (10 December 2010) The future of LinkedIn and the economic graph. LinkedIn.com: http://www.linkedin.com/today/post/article/20121210053039-22330283-the-future-of-linkedin-and-the-economic-graph [accessed 23.04.14].

2   Crabtree, Steve (8 October 2013) Worldwide, 13% of employees are engaged at work. Gallup World: http://www.gallup.com/poll/165269/worldwide-employees-engaged-work.aspx [accessed 23.04.14].

3   Clifton, Jim (19 June 2013) Millions of bad managers are killing America's growth. The Chairman's blog, Gallup: http://thechairmansblog.gallup.com/2013/06/millions-of-bad-managers-are-killing.html [accessed 23.04.14].

4   Goudreau, Jenna (25 February 2013) Back to the stone age? New Yahoo CEO Marissa Mayer bans working from home. Forbes: http://www.forbes.com/sites/jennagoudreau/2013/02/25/back-to-the-stone-age-new-yahoo-ceo-marissa-mayer-bans-working-from-home [accessed 23.04.14].

Preface

Imagine you were alive in France in 1492 when you heard that an explorer, Christopher Columbus, had discovered what would become known as the 'New World' of the Americas. At that time, the 'world' as you understood it comprised only Europe, Africa and Asia. His voyage was not the first but it was the most significant and, from the moment Columbus stepped onto land in the Bahamas, your world view was recast. What you had once believed to be true no longer was, and railing against change was destined to fail. Columbus was just one of the many individuals whose endeavours artistic, scientific, religious and commercial would shape an extraordinary period we call 'The Renaissance'.

Today we are amazed, inspired and sometimes shocked by the scale of scientific discoveries, economic and social upheavals, and technological innovations we hear about on a daily basis. All the music you have ever wanted right there in your hand; the entirety of written human knowledge available via that small device; your key medical indicators visible thanks to the same tiny circuitry; explorations into deep space that chart the moment our universe appeared; photos we take of ourselves that generate millions for charity in days. The scale of innovation and exploration is staggering.

And yet, while each invention takes our breath away, it is the sweeping arc of ideas and insights that really moves us as human beings. We each know that we are challenged as never before environmentally, financially and socially, but at the same time we are enabled to see and therefore comprehend how these struggles are shared across our planet. We are alive during a remarkable era in human history; a period of perhaps several decades that will shape not only this century but many that follow. We are living through a Digital Renaissance.

This Digital Renaissance affects all aspects of life, but the focus for myself and my co-author Elizabeth Marsh is on 'work', as this is our area of understanding. How does this birth of creativity, innovation and breakthrough affect the ways in which each of us works? What is its value for organizations of all sizes? In 2012, I wrote *The Digital Workplace: How technology is liberating work* to define and explore a new term 'the digital workplace' that I had coined to chart the emerging digital world of work (as opposed to 'the physical workplace' with which we were so familiar). It has been rewarding to the cities where most people still live. The effect of the Industrial Revolution was a hugely productive phase for nations but a destructive and cruel period for work itself. In the nineteenth century, Charles Dickens chronicled the pain of the factory; then offices took over as the dreary and soul-destroying places managers and administrators would trudge into day in, day out. Certainly, we still have no shortage of archaic companies (and managers within them), who insist on outmoded work formats, demanding that staff come to a fixed physical space each day. The Google culture, for instance, may be inspiring in many ways, but its insistence on face-time in large offices every day is a straitjacket for staff who expect to be able to choose how and where they work.

The exciting development though is that this third major transformation is beginning to liberate work and to usher in an era in which everyone in work has increased capacity to derive far more reward (beyond financial) from their daily endeavour. What is so inspiring to see is that this increase in fulfilment from work does not simply apply to purely managerial or knowledge-based roles, but also to many more repetitive roles undertaken by workers in call centres, factories, delivery operations and warehouses, where digital empowerment (as we have seen with Barclays) can create a dramatically better experience of work. In turn, this can help to stimulate innovation, drive competitive advantage and bring other fundamental organizational benefits, which ensue when we are fully engaged in our work. This book charts the scope and scale of this digital transformation of work and how to implement essential changes inside your own work and organizations.

How Zara Does It

The implications of the Digital Renaissance of Work are far-reaching, stretching beyond the individual worker to company processes and culture. One example illustrating this release from tedium and the resulting wider business effects can be found in successful clothing store Zara, which has become the world's largest fashion retailer. One of the reasons Zara has been able to trump its competitors is through the digital design of its work processes, which embed digital connections into all parts of the staff experience. Through this, Zara is able not only to respond to customer trends and tastes quicker than everybody else but also to feed its young consumers' appetite for a continual supply of new designs; the company can produce a new garment in ten to 15 days from design.[8]

Mobile technology deployed within the retail units is very important and Zara has been an early adopter, using mobile devices in one form or another since the 1990s. Sales assistants have had personal digital assistants (PDAs) since the early 2000s, allowing them to immediately transmit customer reactions to sales and to record opinions. When a sales assistant asks a customer a relevant question about their experience, the responses can be recorded in real-time. Supplemented by real sales data, this allows Zara to develop powerful insights and to respond rapidly with a constant stream of new designs in store. Gone are the days of the bored sales assistant mindlessly enquiring 'Can I help you?' Instead we find the new digitally knowledgeable retailer who can connect the Zara marketplace with production and witness the difference on the shelves within weeks.[9, 10, 11]

Now, I can hear the protestations from those who counter: 'But my job is still awful … What about the street-sweeper? Many people hate their work and would retire tomorrow … that's why they buy a lottery ticket each week. Just who are you kidding? Work as enjoyment, give me a break!' Of course I don't believe that every working role has somehow miraculously been transformed overnight into a paradise previously reserved for Picasso or Moby, but more that a digital groundswell of change is happening and beginning to unstitch the way work is structured and how it is performed – and doing so relentlessly and for the better.

Even in situations that many would regard as some of the toughest working conditions in the world, and where the benefits of technology would seem least likely to be felt, the digital workplace is helping to drive change for the better. For example, UK high street retail institution M&S is now using mobile technology to directly survey all the employees who work in its supply chain within India and Sri Lanka, to ensure that working conditions and employee satisfaction adhere to standards the company and its customers would want to see.[12]

As the Digital Renaissance of work evolves, the impact waves of the digital workplace will crash constantly on the shores of every organization, both large and small, from 500,000-person Deutsche Post/DHL to one-man band AAA Plumbing and Heating Services, creating both huge challenges and beneficial effects for those who choose to embrace the changes. The trend for staff using their own devices for work (so-called Bring Your Own Device; BYOD) is just an early tide flowing in and resistance is futile.

The Digital Journeys Addison Lee Makes

Take, for example, London-based cab company, Addison Lee, which is a digital tour de force in the private hire car industry. There will be more on this later but, as a taster, just one aspect is that their call centre, like Mark Warner, is not one centralized office space; this operation includes the ability for those handling the calls to work from home, using a flexible shift pattern system. For the customer, the experience is identical to calling the main centre but for the home-based staff the option to pick and choose the design of their working day produces happier, more committed and productive staff – even though the task of answering the calls has not inherently become any more interesting.[13]

urging their graphics teams to come up with yet more evocative ways of visualizing networks, data flows and viral systems, often resorting to shooting spears of coloured lights or a planet swirling with data.

In the same way, we each struggle to grasp the digital worlds where we spend ever-increasing amounts of our day, work and lives. Where are we when we are logged into Facebook? What can we touch when we are with five colleagues in a virtual Webex meeting? The challenge for the TV stations echoes the issue we all face in the digital workplace: we spend so much of our working day in digital worlds, but these are invisible while at the same time very present.

A New 'Digital Geography'

This digital realm of work is a new landscape we inhabit. But do we enter these digital worlds because we want to or because we have to? If they existed in the physical world, would they be examples of those gardens and homes we choose to visit or live in – or are they places we would make sure to avoid? Are these digital worlds worth working in?

So far, the digital realms we inhabit so much have generally been disappointing. We are impressed by their capability because they are still so new and innovative, but we enter them because we must do so in order to live and work rather than because we enjoy them as we would a beautiful garden or even a pleasant, light-filled refreshed office building. Most senior digital executives know this is a problem. From my conversations, it is clear that most CIOs believe that their employees want, and would gain individually and collectively from, increased access to 'enterprise applications' – intranet, mobile, HR, unified communications, shared documents, SharePoint and so on – but equally these CIOs tell me that such systems are often too complex and integration is expensive. They recognize the challenge but frequently struggle to find a remedy.

Based on the fragmentation my colleagues and I in the Digital Workplace Group (DWG) observe in organizations each day, if your digital workplace were a bricks-and-mortar building, the chances are it would be condemned right away on health and safety grounds. Once you've got online, you follow a link, only to find you are being prompted for your password for the umpteenth time. What was it again? Now it's time to leave the office to go to a client … OK, I can review the meeting notes in the taxi … but hang on, I can't access them on my iPad …

Padmasree Warrior, Chief Technology and Strategy Officer of Cisco, said in 2013 that a tiny 1 per cent of the connection potential of the digital world has so far been realized.[2] To her 1.5 million Twitter followers, what this statement conveyed was that we are still in the very earliest stages of shaping and designing the digital worlds we work and live in. In physical world analogies, we are still living in caves.

Digital Worlds Where We Want to Work

So, given the criticality of the digital workplace, we need to start creating digital worlds that attract and thrill us, just as we have in the physical world over millennia. The digital workplace (this realm we dive into each day) must become beautiful, pleasurable and productive. It is an insistent and compelling world but we need to start regarding it in the same way as we have the physical workplace; as a place we can design and craft into somewhere we want to inhabit, rather than merely somewhere we are obliged to enter.

Within the consumer technology space we have already started to experience quality and aesthetics, for example, in Apple products and interfaces; but, for so many, the digital workplace remains a thoroughly underwhelming experience. As the author Peter Hinssen humorously referred to it, work is now 'the brief period during the day when I use old technology'.[3] The opportunity is huge and it may be that this new connected world will offer a new stage in human evolution when, as a species, we learn how to live and work in digital realms, evolving capabilities and the dexterity needed for the worlds we will explore digitally. So much media focus is on the dangers of this digital universe: the effect on children, the dark web, cyber threats, and so on, but while these are all natural and important concerns, the flip side that this unique, connected space offers, is a rich, beautiful world, unconstrained by physical limitation. Time and geography collapse and a freedom is possible that, as Padmasree Warrior says, we have only just begun to explore.[4] This is a seemingly infinite digital universe that we can and will colonize as a species, just as we have our physical world.

But, so far, the reality of the digital workplace is one of fragmentation and huge variations in quality and access. Most digital workplaces provided by enterprises have very poor user experience, with interfaces that are not intuitive and passwords that must be constantly re-entered as we move between services and devices.

Bouygues Telecom's Worlds Worth Working In

One company that has created an effective digital workplace though is Bouygues Telecom, a French mobile telecommunications firm and Internet service provider. Bouygues Telecom has created 'B.desk', a highly immersive mobile digital workplace, which is optimized for tablet use and touch screens. B.desk provides access to the firm's existing SharePoint intranet, social network, video channels and SharePoint document libraries through an individual's own personal tablet or smartphone, or through a corporate-owned device. It also uses a private cloud facility, which allows documents to be effectively synchronized between devices, as well as with services accessed through desktop or laptop devices.

This effectively means that whatever you can do on the desktop you can also do on a mobile device. This is not only valuable for the company's many employees who are 'on the go', such as engineers and sales staff, but also tablets are increasingly being seen on people's desks with B.desk open because of the quality of the user experience. Although the system was initially developed for internal use, it has now been prepared for commercial release so that it can be extended to sister companies within the Bouygues Group and also to Bouygues Telecom clients.[5]

When it comes to creating digital worlds worth working in, the motivation is far from just a question of digital aesthetics. As the Bouygues Telecom story shows, the power of success creates value for customers and, in this case, even has commercial value elsewhere within the group. And, if achievement in this area can be demonstrated to add value, there are also severe downsides to tolerating the lamentable status quo. Research by DWG[6] has shown that poor digital workplaces result in: low levels of service adoption (in contrast to the high engagement levels enjoyed by Barclays staff);[7] loss of productivity (a Kellogg School of Management study reports that better digital connectivity improves the ability to find information by 31 per cent);[8] erosion of employee satisfaction; struggles to recruit and retain staff; and an increase in security risks through human error.

How One Coffee Chain Connected

The right digital workplace can engage and motivate staff. Heine Brothers, a small Kentucky-based chain of 14 coffee shops, found that it was getting an open rate of just 20 per cent for emails sent to the personal email accounts of their 240 generally young, hourly paid staff, who had no access to company technology. Given that all staff had their own mobile phones, Heine Brothers decided to use an app, called 'Red E App', which allows staff to subscribe to the company network. Once verified, the company can communicate messages via the app and now achieves 98 per cent open rates. The app was, for instance, effectively used when a problem arose with issuing the staff's pay cheques. The company was able to alert the employees to the issue and reassure them before they had even started to worry about what might have gone wrong.[9] This simple improvement in the digital workplace has built culture and retention, and the impact (as at Barclays) is dramatic rather than a small improvement.

The realization that the new digital world is not just a technical or functional space, but more akin to the physical world, emanated from our first 'Digital Renaissance Man' – Steve Jobs. Jobs knew from his early days as a calligrapher that beauty makes us feel good, gives us pleasure and, crucially, that this love for what we experience can drive vast commercial returns. Apple designed beautiful digital environments and, in the early days of the iPod, iPhone and iPad, this was unique. It is sweetly ironic that Steve Jobs, who had no interest himself in the enterprise space, brought this digital delight into companies as the benchmark for digital aesthetics.

The enterprise apps revolution has energized companies like British Airways, Starbucks and Delta Air Lines to join the model of agile start-ups such as Dropbox, Evernote and Flipboard, and to start developing digital services for the mobile devices their staff carry with them all the time. Smart thinking since, as Gartner's Peter Sondergaard estimates, by 2020 there will be up to 30 billion devices (mostly mobile) connected to the Internet, while others say this figure is conservative bearing in mind the arrival of mobile technology embedded into everyday living and working components.[10] What these companies are realizing is that they must become rapid developers of digital workplace technologies that fit their culture, patterns and structure, or the fragmented digital landscape will hamper productivity, engagement and the flow of work. This approach becomes essential in a company such as JetBlue Airways, where not only do 6,000 of the 8,000 reservation agents work from home but also many of the younger staff are demanding to use their own preferred technology in their work.[11] Given that Juniper Research says the number of employee-owned smartphones and tablets used for work will exceed 1 billion by 2018, the quality of this digital world increasingly becomes more fundamental to organizations than that of the physical workplace.[12]

Like Being in the Same Room – Physically

Where to start in tackling this fragmented digital workplace challenge? During a talk at Adobe's HQ in San Jose, I was asked where the company should focus its digital workplace resource, given that its staff want 'everything we do to look and feel perfect'? Knowing that Adobe already has one of the better-looking digital workplaces I've encountered, my answer was to 'focus on the quality of your real-time audio and video connections because progress there will render many physical meetings unnecessary and enable new levels of digital collaboration globally', very important for entities like Adobe with staff situated worldwide. Certainly, telepresence facilities are now normal within the enterprise but the race is on to extend the technology to deliver more personal, immersive and flexible telepresence solutions. Some of this has resulted from an extension of the current technology (for example, in enabling direct 'eye-to-eye' contact in a telepresence environment) but other innovations are completely redefining the telepresence experience.

One of the companies that has been researching in the telepresence space is Microsoft. Internally they have a vision called the 'Magic Window', a series of life-size visual collaboration spaces for remote parties, often in 3D, with augmented reality information to complement it.[13] Microsoft is actively creating various prototypes that provide telepresence and also capture gesture-related information.

One near-realization of the Magic Window is the Holoreflector, which combines a huge screen with a mirror behind it with a Kinect sensor and Windows phone to track movement. A more practical application is the smaller IllumiShare, which looks rather like a desk lamp but combines a camera and projector, allowing individuals to share a common illuminated area remotely, on which they can write, draw, design collaboratively or show objects.[14] The company has also created the IllumiRoom, which projects images beyond a television screen and can be used for enhanced entertainment.[15] The Head of Research at Microsoft, Peter Lee, is encouraging more research into telepresence, for example, challenging his team to tackle questions such as 'Would it be possible for musicians to collaborate through time and space?' and 'What would we learn by enabling children to really engage at a distance?'[16]

Another direction for telepresence is using robots to allow a moving connection to enable a remote presence within factory and distribution situations. Here a robot on wheels, which can be controlled by a remote user via their laptop, houses a camera, thereby enabling them to view the facility. In return, the telepresence robot displays a screen connecting with a webcam of that user, so some form of remote interaction can take place. These robots are now commercially available to buy or rent with the possibility of a test-drive over the Internet, a slightly eerie but enthralling experience that is worth experiencing.[17]

Inside Out

Along with the engagement and productivity gains to be made from a compelling digital workplace comes the just as important avoidance of the digital disillusionment and disappointment that afflict many organizations – and the cost of this is real. One well-known online insurance site had a terrific external brand and user experience and, attracted by that, staff would join, only to fire up the technology on day one in their new job and feel their hearts sink. They found themselves thrown back five years into a digital dark age as the system clunked and shuddered into action. One new hire simply left at lunchtime and others were back on LinkedIn that evening for sure.

Research by Deloitte in Australia suggests that 9 per cent of workers without flexible IT policies, such as BYOD and access to social media (both potentially important components of the digital workplace), were planning to leave their workplace. This compared to 6 per cent planning to leave who could access flexible IT policies. This difference might sound small, but the differential is highly significant across a large organization, especially when translated into employee retention rates.[18]

However, the value in progress touches not only the staff but will affect customers and the bottom line as well. Changes in how the company works digitally inside the organization will also have huge impacts on the customer experience externally. An excellent example is the aforementioned London-based cab company Addison Lee, which has grown from a one-car operation founded in 1975 to a 3,500-strong fleet in 2013, carrying about 10 million passengers a year. In fact, the growth has been so rapid that the company was sold in 2013 to private equity group Carlyle Group for £300 million.[19]

Digital technologies have been integral to the company's success, not only in the ability to effortlessly scale operations, but also in delivering the customer experience. At the heart of this is the Auto Allocator, which analyses all the bookings for the 3,500-vehicle fleet and then allocates a cab for each customer based on location, priority and the position of the cabs. Empty vehicles are repositioned to potentially busy areas and drivers can also request jobs at the end of their shift on their home routes.[20]

This means that 98 per cent of all bookings are allocated without human involvement and a driver can be instructed within 30 seconds of the booking being made. Using GPS data enables an SMS to be sent to the customer to inform them when the cab is leaving, an estimated time of arrival and then another message when the cab is just about to get to them. The efficiencies gained from this system have resulted in a massive 18,000 miles of 'dead mileage' being saved each day. The Auto Allocator improves operations, customer experiences and reduces carbon emissions.

The company has also used further technology to enhance the way it interacts with customers. An iPhone app introduced in 2010 allows customers to book a cab without having to give their pick-up address, instead using the phone's geo-location facility to locate the customer. This app has been so successful that, in the first six months of 2011, iPhone app bookings alone added 20 per cent to the firm's bottom line.[21] A quarter of all bookings now come through the app.[22]

If we are today working in 'digital caves', utilizing just 1 per cent of the potential future power of the digital world, then setting a digital workplace vision for an organization is essential. We need 'design standards' (equivalent to the architectural and building design standards of the physical world), covering the benchmarks of digital flow and coherence for digital services used by staff, suppliers, partners, customers and the wider marketplace. Imagine what an 11 year old of today will expect when they enter your organization in a decade from now? What I notice in the successful examples so far is the combination of a clear leadership vision of digital excellence, alignment of senior stakeholders, energetic deployment and the patience to allow time for changes to become embedded. The longer the laggards wait though, the further behind they will fall, making the ground they must recover a huge or impossible task.

Notes

1   Alexander, Christopher (1979) *The Timeless Way of Building*. New York: Oxford University Press.
2   Kirkland, Rik (May 2013) Connecting Everything: A conversation with Cicso's Padmasree Warrior; Interview by Rik Kirkland. McKinsey & Co Insights & Publications:: http://www.mckinsey. com/insights/high_tech_telecoms_internet/connecting_everything_a_conversation_with_ciscos_ padmasree_warrior [accessed 29.03.14].
3   Hinssen, Peter (2010) The New Normal: Explore the limits of the digital world. *Across Technology*, p. 17.
4   As 2 above.
5   Step Two Designs (2013) Bouygues Telecom case study. In: Intranet Innovations 2013 report: http:// www.steptwo.com.au/columntwo/announcing-winners-2013-intranet-innovation-awards [accessed 29.03.14].
6   Tubb, Chris (2013) Digital Workplace User Experience: Designing for a flexible workforce. Digital Workplace Group: http://www.digitalworkplacegroup.com/resources/download-reports/digital- workplace-user-experience [accessed 29.03.14].
7   DWG (2014) Barclays Bank 'MyZone' case study. In: Bynghall, Steve (2014) Success with Enterprise Mobile: How tools for frontline employees drive value. Digital Workplace Group: http://www. digitalworkplacegroup.com/resources/download-reports/success-with-enterprise-mobile [accessed 29.03.14].
8   Kellogg School of Management (3 June 2013) The Coworker Network: How companies can use social networks to learn who knows what. Kellogg Insight: http://insight.kellogg.northwestern.edu/article/ the_coworker_network [accessed 29.03.14].
9   Gohring, Nancy (24 July 2013) How do you reach employees who never read email? This company figured it out. Tales from the Cloud, CITEWorld: http://www.citeworld.com/mobile/22170/email- ignored-problem-fixed [accessed 29.03.14].
10 Gartner (11 November 2013) Gartner says personal worlds and the Internet of everything are colliding to create new markets. Gartner press release: http://www.gartner.com/newsroom/id/2621015 [accessed 29.03.14].
11 Belissent, Jennifer (6 June 2013) Forrester: CIOs need to support the workplace of the future now. *Computer Weekly*: http://www.computerweekly.com/news/2240185321/Forrester-CIOs-need-to-support- the-workplaces-of-the-future-Now [accessed 29.03.14].
12 Bhas, Nitin (19 November 2013) BYOD trend drives number of consumer owned mobile devices used at work to exceed 1bn by 2018, finds Juniper Research. Juniper Research press release: http://www. juniperresearch.com/viewpressrelease.php?pr=413 [accessed 29.03.14].
13 As 12 above.
14 Lichtman, Howard (1 March 2012) Microsoft research shows off its next generation of telepresence technology. Telepresence Options: http://www.telepresenceoptions.com/2012/03/microsoft_research_ shows_off_i [accessed 29.03.14].
15 Co, Alex (10 January 2013) Microsoft IllumiRoom projects images beyond your TV. Telepresence Options: http://www.telepresenceoptions.com/2013/01/microsofts_illumiroom_projects [accessed 29.03.14].

16   Buderi, Robert (14 November 2013) The Future of Microsoft Research: One on one with new boss Peter Lee. Xconomy: http://www.xconomy.com/national/2013/11/14/future-microsoft-research-one-one-new-boss-peter-lee/?single_page=true [accessed 29.03.14].

17   Anybots (undated) Anybots virtual presence systems. It's you, anywhere… Anybots corporate website: http://www.anybots.com [accessed 29.03.14].

18   Deloitte (2013) The Connected Workplace: War for talent in the digital economy. Deloitte Access Economics research report: http://www.deloitte.com/assets/Dcom-Australia/Local%20Assets/Documents/Services/Corporate%20Finance/Access%20Economics/Deloitte_The_Connected_Workplace_2013.pdf [accessed 29.03.14].

19   MSN Money (21 April 2013) Cab firm Addison Lee sold for £300m. MSN Money: http://money.uk.msn.com/cab-firm-addison-lee-sold-for-%C2%A3300m [accessed 29.03.14].

20   Addison Lee (undated) Customer-focused Technology: Auto Allocator. Addison Lee corporate website: http://www.addisonlee.com/discover/technology/auto-allocator [accessed 29.03.14].

21   Hall, Kathleen (14 June 2011) iPhone app boosts Addison Lee business by 20% in just six months. Computer Weekly: http://www.computerweekly.com/news/2240104764/iPhone-app-boosts-Addison-Lee-business-by-20-in-just-six-months [accessed 29.03.14].

22   Curtis, Sophie (26 October 2012) Addison Lee's iPhone app revenue doubles in a year. CIO: http://www.cio.co.uk/news/strategy/addison-lees-iphone-app-revenue-doubles-in-year [accessed 29.03.14].

Chapter 3

Where Will We Work in the Physical World?

The office during the day has become the last place people want to be when they really want to get work done. In fact, offices have become interruption factories.
Jason Fried and David Heinemeier Hansson, Remote: Office Not Required [1]

Since I first began talking about the digital workplace in 2009, one question that has come up regularly is 'Where will employees and contractors work physically?' In other words, what is the future of 'the office'? The assumption (wrongly) was that the potency of the digital world of work would force people out of the fixed locations they were familiar with into the cold, never to see each other again. My counter was that the digital workplace is present *everywhere*, including offices, warehouses, retail outlets and factories, so there is no inherent reason why people should vacate their offices. The point is that the digital workplace brings options and choices for the first time.

In reality, since the digital workplace creates alternatives for when and where we work, many people now choose non-office settings for some or all of their working week. Corporate office use in the US fell from 80 per cent in 2010 to 72 per cent in 2012, and even highly social countries, such as Brazil, experience only 61 per cent of workers going to an office four to five times a week.[2] It is not part of the mission of the digital workplace to render offices obsolete or to subtly create nations of home-workers, but some of the hot debates have pivoted around the future for the office, steered nicely by real estate teams tasked with reducing 30 per cent of the cost from the huge sums spent on buildings. Many companies have 'new ways of working' or 'future workplace' initiatives, ideally comprising real estate, human resources, technology and communications, but most have struggled to achieve the right balance between the digital workplace and time spent physically with colleagues (Unilever's Agile Working and Microsoft's Workplace Advantage programmes[3] being notable and more successful exceptions).

programme of any federal agency. In 2012, the *Washington Post* described it as having the 'gold-standard' for a 'government work-at-home program'.[12]

Agency-wide, USPTO now has 9,032 employees working from home between one and five days a week, equating to 83.96 per cent of eligible positions teleworking. Overall, around two-thirds of agency staff telework in some form or another. Overall benefits have included significant cost savings and avoidance, primarily gained through avoiding the need to acquire additional real estate when bringing in new hires. Other measurable outcomes include increased productivity (as a result of fewer distractions) and enabling continuity of operations planning.[13]

As another bonus, telework has helped position the agency as an employer of choice. In 2013, USPTO was ranked first (of 300 federal agency subcomponents) in an annual survey of the US Federal Government's 'Best Places to Work'. The survey is taken by 700,000 employees from 371 federal agencies.[14] USPTO's Commissioner for Patents, Margaret A (Peggy) Focarino, stated: 'This award is a tremendous tribute to the tireless dedication of our hardworking employees, unions, and agency leaders.'

USPTO was an early adopter of telework. As early as 1997, the USPTO's trademark organization initiated a small but successful telework pilot. This longevity means that the programme is now effectively hard-baked into the organization's strategy, enjoying wide support and acceptance covered later in Chapter 13.

The Everywhere/Anywhere Office

So, should we close down all offices now? No, but every company, both large and small, needs to plan carefully for the spaces it expects its people to work in, whether these are owned, leased, rented by the day or hour, or shared with others. Inevitably, if workplace choice is just 'table stakes', as a Deloitte HR executive described it to me recently, then those who make a choice on where and how they work (which will be most staff aside from a small number in manufacturing plants or warehouses, doubtless working alongside an ever-growing robotic population), then fewer people will be visiting central office locations each day. They will work increasingly from properly customized home-working spaces, a far cry from the first-generation 'corner of the living room'. We will see ever more garden offices, new-build homes equipped with a semi-separate work space, and pop-up home offices embedded with all the tech we need, springing up. Staff and contractors travelling to any fixed place of work will become entirely optional in the coming years.

At the same time, what we require when we do meet up with our colleagues physically is as short a journey from home as possible. Major organizations will therefore need far more plentiful physical locations, situated closer to where people live, and possibly shared with other large companies and local businesses. A Shell employee may well want some time physically with his or her colleagues in the company, but do they really want to have to travel to Utrecht or London in order to achieve this? Wouldn't it be preferable to walk or drive into a town or village, or urban neighbourhood, close to where they live? Why are those Googlers being pushed out of central San Francisco each day on Google buses (admittedly equipped with good WiFi) to travel an hour or more each way to Mountain View, when they could sit with a dozen colleagues in a Google café in San Francisco's Mission District? Managers who insist that their teams come to a fixed location each day will decline in numbers. They will retire, receive redundancy, or just lose staff. Choice, options and digital agility will become normal for all organizations, while new younger hires will increasingly select companies with a working culture that feels natural, human and respectful – with choice being paramount.

At DWG we embraced this idea that the office is no longer an essential when we closed our London office in early 2011. Rather than finding we missed having an office, the move has had a positive effect across the company since everyone is now equal in a digital world – liberated from the illusion that our single office was somehow a de facto HQ, where power resided.

Different DWG team members enjoy a range of working styles. Editor Suzanne Clark is a good example: aside from working for DWG, she runs an eco-tours business from her home in Grenada in the Caribbean. Or there's benchmarker Louise Kennedy, who regularly flits between the UK and her mountain climbing base in France. Communications Manager Ephraim Freed is usually based in Seattle, but sometimes he is in London. Producer Skye Crawford works out of the UK, but when she makes regular extended visits to her original home Australia, most people think she is still working in England. All these people carry out work for DWG independent of location. This has never been an issue for our 70 staff and contractors, the majority of whom are based in Europe and the US. Some of these people work regularly for DWG, some work irregularly, and others enter and leave for specific projects.

The lack of disruption involved when DWG shut its office made us realize that most of the firm's interactions were already made online, often via conferencing and collaboration tools like Skype, Webex and Google Drive. Being an officeless company hasn't meant that the company has divested traditional reporting lines, meetings or processes though; nor does it mean that face-to-face interaction is not important. In the

UK we have a monthly physical team day; consultants are encouraged to see clients and members face to face; and management team members from the UK regularly fly over to the US and vice versa. But it has brought into focus the question 'How important is a fixed location office?' For DWG, the answer is that a fixed location has no importance at all.

There Must Be a 'Third Place'

But DWG is a small company with a wide reach and what suits us will not work for all. Certainly, we will continue to see plenty of alternatives springing up that are neither home nor the main office but a 'third place' that suits whatever you fancy on that day. Quiet space, social time with colleagues, views we enjoy – the list will grow as solo-living urban dwellers look for places to work that are near where they live and close to transport hubs. It is great to see established providers like Regus opening drop-in workplaces in railway and motorway service stations, and in local neighbourhoods, with swipe cards provided for staff by their employers. Wherever we go, we need decent strength WiFi, power and now-and-then printing facilities, so Starbucks' days as the go-to 'third place' are limited rather than a durable option.

Recently, attention has been turned to this notion of the third place for work – flexible and convenient working spaces that come with neither the costs nor the inconvenience of the fixed-office location, but which improve on the limitations of working from home and are able to serve the needs of a mobile workforce with unpredictable movements. Third places have emerged in the form of co-working spaces such as Impact Hub,[15] providing amenities and a sense of community for freelancers and smaller start-ups. Meanwhile, in another recent development, a few cafés such as Russia's innovative Ziferblat chain, are operating on an entirely new model, providing WiFi and charging for time spent in the café (3p per minute in the new London outlet) rather than for the coffee and cake consumed.[16] However, these spaces are often not secure enough to serve the needs of employees from larger corporations; they frequently have a membership model; and they are usually located in city centres, so are not suitable for mobile workers travelling long distances who need a place to work on the fly.

Regus has been delivering business centres, managed offices and flexible workspaces since the late 1980s and has now started to experiment with third places designed specifically for the enterprise mobile worker. These are situated in different popular places frequented by travelling workers and usually have the sort of drop-in business facilities often found in hotels. For example, in the UK, the company has signed an agreement with the Extra Motorway Service Areas Group to open business facilities at motorway service stations in the south-east. It is also experimenting by opening facilities in four Staples stores across the UK.[17] In Berlin, Germany, Regus has partnered with oil company Shell to provide Regus facilities at 70 different service stations across the city. Facilities range from WiFi to docking stations to business lounges.[18]

Offices Fighting Back

Disruptive change is not a linear process though and it is fascinating to see how the offices themselves are waging a campaign to attract people back. This is an exciting development with potential to surprise us, as physical workplaces are reimagined. We are attracted to places that can give us something special and, according to global architectural practice Arup,[19] employees will not come to work just in order to meet their colleagues – after all, how many people can you usefully meet in any given day? Research shows that colleagues more than 100 metres away might just as well be 100 miles away, as you will almost never talk to them.

Arup in its 'Living Workplace' report (2012) suggests that new-style workplaces will have to design social, gastronomic, health and lifestyle facilities that make the trip to that venue attractive.[20] In DWG we have found that, as we become more at ease in digital worlds, we crave physical interactions that enable meaning and connection at deeper levels, even more so than before. To quote one of my heroes, Timothy Leary (Harvard professor and author) from the 1990s: 'In the future, physical meetings will become sacred, taking on mythic importance.'

Jason Fried and David Heinemeier Hansson make a similar observation in their recent book *Remote: Office not required*:

> *By rationing in-person meetings, their stature is elevated to that of a rare treat. They become something to be savored, something special. Dine out every once in a while on those feasts and sustain yourself in the interim on the conversation 'snacks' that technology makes possible.*[21]

The locations companies own, lease, rent, share and so on, will need to feature enhanced technology and in-person experiences you cannot get elsewhere: superior unified communications, better food, innovation areas showcasing the company's R&D, integrated services like dry cleaning, shopping and child care, so that it becomes more like a hybrid shopping and civic centre than a corporate facility.

That said, the overall pattern suggests a shift in power. Historically, the office took precedence and was the main driver in how we worked. Technology, initially in the form of desktop PCs, was added to the office but the office itself reigned supreme. Now it is the digital workplace that leads and physical workplaces are having to adapt in order to contribute whatever the employer decides the digital cannot provide. For example, consulting firm Accenture now regards physical workplaces as 'strategically irrelevant', since their 260,000 people can work anywhere.[22] Physical workplaces will play second fiddle to the digital and those in corporate real estate will find their industry remains useful but of marginal value only.

Different Teams – Different Designs

But the good news is that different teams will choose varying 'physically together/alone digital' designs, based on what they do and their individual needs. Field engineers might meet once a week locally. Developers may work together two days a week. Salespeople could spend weeks without ever meeting physically. The C-suite will design their own weeks based on their preferences and likely have two days a week together maximum but, like everyone else, require persistent digital presence and connection wherever they are. What is a fact is that, while we will travel for colleague and workplace cultural connection, we will not want to travel far. We would ideally like to walk or take a short ride close to where we live – and this means that work-enabled facilities will need to be located closer to where the distributed workforce lives. Smaller and far more numerous workplace options are the inevitable consequence.

One other reflection based on DWG's experience of not having a physical location is that it appears the lack of an office is actually strengthening connection and collaboration. Everyone makes their presence felt digitally. When recently one researcher in the UK was suffering after her partner was involved in a car accident, the team around the world reacted at once through Yammer, phone and instant messaging (IM), and she remarked at how supported by the messages she instantly felt. Other colleagues examined her diary and shifted activities into their own names, releasing her from the work as best they could. This happened within hours. She was also able to continue with as much work as she wished because she could do it while tending to her partner at the hospital. She redesigned her week and we all (including her) benefitted. In a physical office, colleagues would naturally have rallied round also, but less effectively because it would have been based around just those who happened to be working there that day. The digital workplace was faster, more practical and included every colleague, no matter where they were when news of the problem surfaced.

Notes

1   Fried, Jason and Hansson, David Heinemeier (2013) *Remote: Office not required*. Crown Business.

2   Belissent, Jennifer (6 June 2013) Forrester: CIOs need to support the workplace of the future now. *Computer Weekly*: http://www.computerweekly.com/news/2240185321/Forrester-CIOs-need-to-support-the-workplaces-of-the-future-Now [accessed 29.03.14].

3   See Appendix 3, Creating a Better Place to Work: Microsoft's Workplace Advantage programme.

4   Goudreau, Jenna (25 February 2013) Back to the stone age? New Yahoo CEO Marissa Mayer bans working from home. *Forbes*: http://www.forbes.com/sites/jennagoudreau/2013/02/25/back-to-the-stone-age-new-yahoo-ceo-marissa-mayer-bans-working-from-home [accessed 29.03.14].

5   Vance, Ashlee (9 October 2013) At HP, Meg Whitman wants people to show up for work. *Bloomberg Businessweek*: http://www.businessweek.com/articles/2013-10-09/at-hp-meg-whitman-wants-people-to-show-up-for-work [accessed 29.03.14].

6   Grubb, Ben (19 February 2013) Do as we say, not as we do: Googlers don't telecommute. *The Sydney Morning Herald*: http://www.smh.com.au/it-pro/business-it/do-as-we-say-not-as-we-do-googlers-dont-telecommute-20130218-2eo8w.html [accessed 29.03.14].

7   Richtel, Matt (22 October 2011) A Silicon Valley school that doesn't compute. *New York Times*: http://www.nytimes.com/2011/10/23/technology/at-waldorf-school-in-silicon-valley-technology-can-wait.html? [accessed 29.03.14].

8   See Unilever case study in: Bynghall, Steve (2013) Digital Workplace Fundamentals: The integrated approach. Digital Workplace Group: http://www.digitalworkplacegroup.com/resources/download-reports/digital-workplace-fundamentals [accessed 29.03.2014].

9   As 8 above.

10   Herman Miller (undated) Modes of work. Herman Miller living office: http://www.hermanmiller.com/content/dam/hermanmiller/documents/news_events_media/Living_Office_Modes_of_Work.pdf [accessed 29.03.14].

11   Lewis, Katherine (2010) President Obama's speech at Fortune's Most Powerful Women Summit. About.com Working Moms: http://workingmoms.about.com/od/executiveopportunities/a/President-Obama-At-Fortunes-Most-Powerful-Women-Summit.htm [accessed 29.03.14].

12   Rein, Lisa (11 September 2012) Telework guru takes phoning it to new levels at patent office. *The Washington Post*: http://articles.washingtonpost.com/2012-09-10/local/35494462_1_telework-patent-trademark-office [accessed 29.03.14].

13   See interview with Danette Campbell in: Bynghall, Steve (2013) Digital Workplace Fundamentals: The integrated approach. Digital Workplace Group: http://www.digitalworkplacegroup.com/resources/download-reports/digital-workplace-fundamentals [accessed 29.03.2014].

14   USPTO (18 December 2013) U.S. Patent and Trademark Office ranks #1 in Best Places to Work in the Federal Government. USPTO.gov press release: http://www.uspto.gov/news/pr/2013/13-40.jsp [accessed 24.04.14].

15   Impact Hub (2014) What is Impact Hub King's Cross? https://kingscross.impacthub.net [accessed 29.03.14].

16   Boyle, Sian (13 January 2014) Ziferblat, the pay-as-you-go café. *Evening Standard*: http://www.standard.co.uk/goingout/restaurants/ziferblat-the-payasyougo-cafe-9056841.html [accessed 29.03.14].

17   Purdy, Steve (2013) Regus business hubs to launch in Staples. Workplaces Regus blog: http://blog.regus.co.uk/latest-news/regus-business-hubs [accessed 29.03.14].

18   Kemp, Phil (2013) Regus and Shell launch city-wide workhubs in Berlin. Workplaces Regus blog: http://blog.regus.co.uk/latest-news/regus-and-shell-launch-city-wide-workhubs-in-berlin [accessed 29.03.14].

19   Arup (2012) Living Workplace: Delivering the workplace of the future: http://publications.arup.com/Publications/L/Living_Workplace.aspx [accessed 29.03.14].

20   As 19 above.

21   As 1 above.

22   Based on a personal communication.

Chapter 4

Collaboration – and Why 'Teamwork' Needs a Makeover

The lightning spark of thought generated… in the solitary mind awakens its express likeness in another mind.

Philosopher and Writer, Thomas Carlyle[1]

We seldom work alone. At times yes, but generally we work with others. Historically this was called 'teamwork' and a vast industry, involving experts and skills, grew up around the need to build effective teams. Groups of employees would head off for outward bound activities involving a range of ludicrous tasks that were decent fun but achieved nothing much else. I remember spending two days participating in a myriad of odd activities at a hotel and activity centre near London in the 1990s with colleagues, only to afterwards find no lasting value at all from the 'team building' elements. That said, teamwork did matter in work because it was all there was in the neo-digital age.

But today we work constantly with large numbers of different people in multiple locations; many we may never meet but we still interact with them as we share knowledge and complete processes. Our non-work lives also now involve a far wider network of people than ever before through all the 'loose ties' that populate them. We talked earlier of the assistant marketing manager for Dove; her working day cannot be summed up by a simple notion of teamwork. What she does throughout the day is to collaborate almost endlessly: she sends and reads numerous emails, posts regularly on networks such as Chatter and Yammer, accesses team sites where shared documents are refined digitally by colleagues, skims through a stream of IMs, often while simultaneously joining real-time conversations. We all collaborate persistently.

There has been extensive research to show that people are more productive when they work both remotely and collaboratively than they are when in an office – hardly strange given the levels of distraction in an office.[2] Cisco found that 60 per cent of its staff said digital collaboration was far more effective than in-person collaboration and these responses are particularly strong in China, India and Brazil, demonstrating that this shift from the physical to the digital, and from individual to collaborative work, is a universal pattern and not just for the developed countries and their knowledge-workers.

If someone were to ask me whether I was 'in a team' or a 'good team-worker', I really couldn't answer any longer. The reality for me as a small company CEO is that my colleagues and I are collaborating continually as we work. This collaboration is not only internal but also with various external entities and customers, further blurring the concept of 'teamwork'. For me, a typical day involves around 40 emails, 20 posts on Yammer, reading and responding to 50 or more Skype chats, accessing shared Google docs, downloading reports posted that day on Basecamp, and the list continues. Some of these collaborations are with colleagues, while many others are with clients and partners, or through LinkedIn with people I know in the industry and wider marketplace.

The concept of the team and teamwork is a relic of the Industrial Age and needs to be reinvented – given a makeover for the Digital Renaissance. In the digital workplace, we are finding that we need to become adept at digital collaboration, working with people we may never meet in person. We need training, guidance and engagement in this new work format. The skills needed are those to build trust rapidly in a digital world; to manage your reputation constantly; to remain digitally visible to your organization; and to expand your network of connections. Teamwork has become attached in our minds with the 'fixed-ness' of the traditional office but now, in the digital world, we need workers across all fields to acquire the skills of 'digital collaboration', as they learn how to maintain more relationships and constantly extend their communication style.

The Enduring Need for Training and Guidance

When discussing this need for skills in virtual collaboration with a senior HR manager in a large oil company, he explained that the company was 'beyond the stage of being open to training – we have just developed new habits'. But that's not really enough when we consider what is at stake. We spend much of our days collaborating and the only 'training', if we can call it that, has been the time we have spent gaining friends on Facebook and then managing some of the testing waters of that network when we have posted a silly comment or photo and been taken to task for it by our 'friends'.

When Virgin Media deployed its successful collaboration space using Cisco Social, the functionality was easy but the company nevertheless spent time training and guiding people in how to make best use of the digital environment. They did not foster teamwork but focused instead on collaboration skills. Virgin Media also catered for the dizzying array of methods by which employees can communicate, allowing a choice between email, microblogging, video conferencing, text chat or good old-fashioned telephone, from the same tightly integrated environment,[3] of which we will see more in Appendix 4. Coca-Cola Enterprises offers a similar number of collaboration channels, but ran a launch campaign called 'Get Connected', which described the use of each one in the context of business needs and different uses.[4] We need that level of support in order to become adept as digital collaborators if we are to optimize how we work.

My colleagues and I witness the collaboration efforts of so many companies, enabled by some very good, as well as other less powerful, digital workplaces. In many cases, a combination of the simplicity of the technology and the right organizational culture are the main ingredients that result in good collaboration. For example, PwC was able to take Jive software largely out-of-the-box, deploy it globally as a 'social' platform, and achieve almost 100 per cent adoption in some territories. Staff are interacting and collaborating with energy and purpose, and I suspect this is down to the kind of people PwC employs, as well as clever deployment of technology; this is covered more fully in Appendix 1.

Digital DNA at Shell

Oil giant Shell has been a pioneer in collaboration. As a company, it has led the way in sharing knowledge and practice since the 1960s, and this experience has proved invaluable. Working this way has become part of the 'Shell DNA', building a rich heritage of knowledge management (KM) and now digital collaboration. The use of a corporate wiki and enterprise-wide discussion forums has resulted in what KM Lead, Andy Boyd, has described as 'audited savings' of between US$300 to US$400 million per year, as well as a culture of collaboration that drives very high adoption of the various different platforms.[5]

In the late 1990s, the company established global 'communities of practice' to group professionals together across worldwide locations. To facilitate discussion and collaboration, in 1999, Shell invested US$1.5 million in an off-the-shelf system so that employees within the communities could share technical data, ask questions and discuss topics.[6] The technology comprised a simple discussion board and access was open to everyone at Shell. Even in the first year this was estimated to have a value of US$200 million in terms of cost savings and new revenue opportunities.[7]

As the success of the communities grew, the global networks started to be embedded into key company processes and, by 2000, 70 per cent of staff had joined one of the networks. The discussion boards (called SIGN, standing for Shell Information Global Network) are still going strong with, in 2012, 70 active communities, 38,000 active members and around 280 postings per day. In 2005, the KM team at Shell established an enterprise-wide wiki to offer an up-to-date 'internal encyclopedia', which encompasses all sections of the business. It includes technical, non-technical and training material. All employees can edit the wiki and, by 2012, there were more than 7,000 topics. Retirees from Shell are interviewed for content for the wiki and also help to review entries.

Shell continues to innovate around KM with savings of hundreds of millions of pounds envisaged with the introduction of a semantic search facility. This will automatically identify the information needs of engineers and help to reduce an engineer's full training period from nine years to eight years.[8] What the Shell story shows is that effective collaboration requires investment, strategy and patience.

Can I Trust You Virtually?

Building trust in digital worlds is another hot topic that comes up frequently. At a large HR conference in New York, one senior leader stated that his objection to the digital workplace was that building trust requires in-person contact. Without that, he explained, trust is impossible. My reply was that I have met people in person whom I have not trusted at all, while I have met plenty of others digitally whom I have trusted at once. Within DWG, two colleagues separated by the Atlantic Ocean were able to organize a 24-hour live online event but only met face to face once it was underway. The building of trust in person is often exaggerated. Sure, meeting in person at the right moment can help enormously to deepen a relationship, but it is often not necessary in order to build trust. Trust comes more from the ways in which we collaborate and communicate digitally, and from the reputation we have in the mind of the person with whom we are in conversation. In the Digital Renaissance we need (and are successfully developing) new beliefs about trust that are liberated from this requirement to meet in person.

The Connection is 'Automattic'

Shell is a very large organization, but an example of a smaller unique global company is Automattic, the power behind the WordPress platform, which supports more than 132 million unique users per month. Automattic employs just 220 people and (like DWG) does not have a physical headquarters. This is, dare I say it, even more remarkable than DWG, considering that WordPress has more unique users per month than eBay, which employs more than 30,000 people. To make things even more complicated, all their people work from home in a variety of different global locations, although there is a lounge facility in San Francisco.[9]

With such a highly dispersed workforce, collaboration is very important for Automattic and it uses a variety of tools. Much of this centres around the WordPress P2 theme, which turns WordPress into more of a collaborative microblogging tool, although others also utilize channels like Skype. Because most of their collaboration tends to be virtual, it means that when there are physical meet-ups these acquire a new significance. For example, Sara Rosso, VIP Global Services Manager, has said:

> *Most of the best days I have had at Automattic have been during our in-person company meet-ups. You might think that this means we should work together more often but, on the contrary, I think that because we are a distributed company we enjoy our time together in person that much more … The best part is working on a week-long project with people who are not your normal team members and presenting your project at the end of the week to the entire company.*[10]

But collaborating virtually also requires a lot of process and discipline. Rosso explains how the team uses the expression 'to P2 it':

> *It's a verb we use to constantly remind ourselves to document our decisions and discussions no matter where they happen, online or offline. Our internal blogs all use the P2 theme, so it's where we want that information to be shared and documented for the rest of the company.*[11]

As well as what some at Automattic dub 'over-communication', the company has also introduced various processes to ensure smooth collaboration. Firstly, all newcomers work on a contract project prior to joining, to see if they are a good fit. Secondly, if somebody misses a target on a project, a fellow staffer or manager will always reach out to them to see what has happened.[12] Ensuring tight virtual collaboration allows the company to deliver a successful platform while maintaining its unique culture, which results in initiatives such as an open vacation policy with no set days to take per year.[13] But, to get there, employees and contractors need training and 'hand-holding' in the best practices of digital collaboration. Just watch a spoof YouTube scenario of a teleconference 'gone wrong' to understand why the new habits of collaborating digitally do not arise naturally![14]

Remember our Verizon engineer? What was previously an individual and isolated activity performed in the homes of customers, has now become a collaborative activity with other Verizon colleagues, who may be physically far apart but can add services needed for success. When the engineer arrives, he is part of an extensive collaborative experience we hardly notice as a customer. In the near future, that engineer may even be wearing Google Glass, or the equivalent, so that colleagues can effectively see what he is seeing and hearing, or alternatively he can view what another engineer has done to resolve a similar issue. The engineer's knowledge base and connections have expanded, leaving his customers probably only subconsciously aware of how smooth and effortless this experience is.

Notes

1   Carlyle, Thomas (1831) Characteristics, in *The Best Known Works of Thomas Carlyle: Including Sartor Resartus, Heroes and Hero Worship and Characteristics*, Wildside Press, LLC, p 319.
2   Lister, Kate and Hamish, Tom (June 2011) The State of Telework in the US: How individuals, business, and government benefit. Telework Research Network: http://www.workshifting.com/downloads/downloads/Telework-Trends-US.pdf [accessed 29.03.14].
3   DWG (May 2013) Live tour of Virgin Media collaborative environment. In: DW24 online broadcast. Digital Workplace Group: http://www.digitalworkplacegroup.com/news-events/digital-workplace-24/digital-workplace-24-recordings [accessed 29.03.14].
4   Avvio Reply (undated) Digitally enabling employees to get better connected. Avvio Reply case study: http://www.avvioreply.co.uk/our-work/Coca-Cola%20Enterprises/Coca-Cola-Enterprises [accessed 29.03.14].
5   IQPC Knowledge and Information Management for Oil and Gas Aberdeen (April 2011) Conference notes. The Data Room: http://oilit.com/2journal/2article/1104_12.htm [accessed 29.03.14].
6   King, Julia (16 July 2001) Shell strikes knowledge gold. *Computerworld*: http://www.computerworld.com/s/article/61898/Shell_Strikes_Knowledge_Gold [accessed 29.03.14].
7   Haimila, Sandra (19 February 2001) Shell creates communities of practice. *KM World*: http://www.kmworld.com/Articles/News/KM-In-Practice/Shell-creates-communities-of-practice-9986.aspx [accessed 29.03.14].
8   Goodwin, Bill (10 October 2012) Shell plans to save 100s of millions with 'semantic' search. *Computer Weekly*: http://www.computerweekly.com/news/2240164896/Shell-plans-to-save-100s-of-millions-with-semantic-search [accessed 29.03.14].
9   Automattic corporate website (undated) Work with us: http://automattic.com/work-with-us [accessed 29.03.14].
10 McConnell, Chris (15 January 2013) Inside Automattic, the company behind wordpress.com: Interview with Sara Rosso. Dailytekk: http://dailytekk.com/2013/01/15/inside-automattic-the-company-behind-wordpress [accessed 29.03.14].
11 As 10 above.
12 Silverman, Rachel Emma (4 September 2012) Step into the office-less company. *The Wall Street Journal*: http://online.wsj.com/news/articles/SB10000872396390044357190457763175017265211 4 [accessed 29.03.14].
13 As 12 above.
14 Tripp and Tyler (22 January 2104) A Conference Call in Real Life. Tripp and Tyler on YouTube: http://youtu.be/DYu_bGbZiiQ [accessed 29.03.2014].

Chapter 5

There Are No Jobs – But There Is Lots of Work

Nothing will work unless you do.

Author, Maya Angelou[1]

Anxiety (and at times panic) about work is nothing new and we are currently living through the latest phase of intense concern over work – and powerful questions abound. Where are the jobs? Will the machines replace us? Will we be paid enough? What skills matter? By definition, the Digital Renaissance of Work applies to those *with* work … but what about the availability of work itself? *The Second Machine Age* by Erik Brynjolfsson and Andrew McAfee caused a media storm in early 2014 by proposing the increasingly feasible prospect and fear that the robots will steal our jobs.[2] With good reason, since an Oxford University study from 2013 predicts that 47 per cent of all jobs may end up being automated within the next 20 years, and this includes knowledge fields such as legal work, accounting, real estate agency and technical writing.[3]

We are right to be unsettled, as each wave of structural change (agricultural, industrial and now digital) produces fundamentally new formats, models and locations for work. In the Agrarian Age, most people worked on the land, lived close to where they toiled, and travelled little. In 1500, an estimated 75 per cent of the British labour force worked in agriculture.[4] By 1800, that figure had fallen to 35 per cent. We saw in the Opening Ceremony of the London 2012 Olympics what a firestorm of upheaval was unleashed when the machinery of the Industrial Age was pioneering in the northern towns of England. Into a seemingly idyllic but arduous rural existence, arose the 'dark satanic mills' of the early 1800s; the noise, dirt and sheer scale must have been disturbing beyond measure to people used to an entirely more tranquil world. By the time the Industrial Age had settled down in the late 1800s, we were living mainly in cities, working in factories and offices, with a working-week schedule that could not have been envisaged in the late 1700s.

Today, we are living through another profound revolution in work. Like the movements before, the trajectory of change is unstoppable and comes with consequences we desire, along with challenges we must address socially, economically and politically. As we digitize work – embedding technology into each aspect of what we do (and remember we have realized only about 1 per cent of the digital connection potential to date, according to Cisco's Padmasree Warrior)[5] – we delight in the freedoms the Digital Renaissance of Work brings. How wonderful, for instance, to be able to sit on a train and download real-time data for the latest orders successfully delivered to a customer.

So Where Are The Jobs?

But we also react with horror at news of the numbers of young people unable to find jobs and wonder what we are really gaining when 1,500 people are made redundant at a major insurance company because the use of online services has drastically reduced the rate of errors requiring a human to rectify them. Fewer calls mean fewer call centres. For each well-networked young mother able to continue her working life from home for a large accounting firm after her second child is born, there is a newly redundant warehouse operator; the victim of next-generation shelf-stacking machinery, boasting a new interface operated by a central control room manager, which eliminates the need for any human hand ever to touch the goods on the shelf.

Through their earlier book *Race Against the Machine*[6] and now *The Second Machine Age*,[7] Brynjolfsson and McAfee have become spokesmen about the way machines are entering our lives. The question they are most often asked is: 'Where will the jobs be?' The next wave of robots is replacing not only manual labour but professional knowledge-workers as well. Why see an accountant when a machine can complete your tax return faster, cheaper and probably more accurately? *The Economist* in January 2014 said that not only organizations, but whole nations as well, are totally unprepared for almost half of all jobs being automated within the next 20 years.[8] Concepts considered impossible ten years ago, such as driverless cars, are now thought to be inevitable at significant levels in the next decade or sooner, given that the self-driving Google car experiment has already logged more than 300,000 totally error-free miles.[9] This change alone could remove 4.5 million jobs, according to Brynjolfsson and McAfee, as the roles for truck-drivers and taxi-drivers simply vanish. But this is not a one-way street and, while jobs worth having are on the decline, work is being recalibrated in new ways. For example, it has never been easier to launch a start-up, to test propositions cheaply and at scale, and to capture a global market (or ditch the idea altogether and try again with another one), all within months and from anywhere. There are clear signs of another uplifting effect too, as recent articles in papers such as the *Financial Times* report the rising numbers of people in work in the UK, as well as a huge increase in engineering vacancies spanning a wide skills range. The issue is mostly one of a mismatch between the skills required and those available, and this pattern will only intensify unless the education system responds to the profound shifts happening, as we will discuss more fully in Chapter 8.

The Work Redesign

Work is changing shape and yet we still remain fixated on jobs. Employment numbers are big news: jobs created; jobs lost; jobs promised through investment. But jobs, in the sense of five days a week, eight hours a day (wherever in the week these fall), are a format from a disappearing era. Business writer and forecaster Charles Handy coined the phrase the 'portfolio worker' as far back as the early 1990s to predict a pattern whereby people have several streams to their working life rather than just one job they remain in for years.[10] Today in the UK, there is much debate over reports of large numbers of new jobs being created, many of which turn out to be part-time or on 'zero-hours' contracts, offering no guarantee of work in any given week. We have also seen a huge rise in the number of self-employed workers and in small business activity, such that these patterns of work together now account for 60 per cent of private sector jobs and 50 per cent of private sector turnover in the UK.[11] This area is viewed as being far more vibrant than the large companies, which are reducing employee numbers year on year.

The digital workplace is enabling a redesign of work and one fundamental element of this change is the slow, but inevitable, reduction in and dependency on jobs as we have known them. For every position advertised, a wave of CVs arrives in the forlorn hope of navigating the various stages of application and interview to landing the job. In one extreme case, 1,700 applicants applied for eight positions in a branch of a Costa Coffee shop in Nottingham, UK.[12]

Once we have secured a job, we then live with the constant insecurity inherent in employment today. This shift to less secure work is an adjustment that is extremely hard for us to make but the new reality is that the only work security lies in having current, desired skills and knowledge – plus the power of your own networks.

So, there may be fewer traditional jobs but the better news is that, at the same time, there is no shortage of work. As well as a wave of new apprenticeships, the statistics of 4.8 million small and medium-size enterprises (SMEs) in the UK in 2013 is evidence that entrepreneurial and skilled work exists, albeit packaged quite differently.[13] This small-scale economic engine room is made possible by the digital workplace because, for example, any local craftsman who makes home furniture can now, at low cost, access a global market for his 'signature products' through popular sites such as etsy.com, which collates local makers of goods around the world in a perfectly designed website and e-commerce process.

Professional editors working from home, anywhere in the world, can gain global clients like the World Health Organization and process a steady stream of medical volumes at times that suit them with electronic payments for the work received via Malaysia and Brussels. Home-finders in north London can advertise their services by linking their websites to property agencies locally and nationally, so that people looking for a home can gain a hands-on service that prepares shortlists of suitable homes for clients moving to the UK. A small specialist machine tools operator in the US can make innovative mobile apps for internal use and then sell them on to a global market through an app store, helping to establish a global market for its services.[14]

Stretching the Boundaries

Change is also gathering momentum within the more traditional employment market. In my last book, I predicted what I called 'work stretching', where companies of all sizes would employ or contract a broader spectrum of people for roles and projects, ranging from teenagers still at school to those in their 70s and 80s, who are still fit enough to work. This has been happening at quite a pace, and with some celebrated stories such as that of Nick D'Aloisio, who at the age of 15 created a news app, Summly, before selling it to Yahoo! two years later for US$30 million. At the same time, he was hired by Yahoo! (but was only able to work for the company after his school day ended).[15] Similarly, market research group, the Ministry of Culture, has a new paid helper: he is a 13-year-old Max Bryant, who has been featured in a series of videos in which he gives advice about ads for kids. Max is the Ministry's 'Minister of Meaning'. Initially interviewed for market research about snack foods, the company found Max to be so eloquent that it decided to give him his own 'punditry platform'.[16]

At the other end of the scale is home retailer B&Q. Sidney Prior was 77 when his second career as a customer adviser at a B&Q garden centre in Wimbledon began and he finally retired in 2011 at the age of 96.[17] In 2013, one in ten over-65s were still working to some degree, up by 10 per cent from the previous year, a leap of more than 100,000 pensioners still in employment.[18] Certainly, the money often helps, or may quite likely be essential, but often it is now through choice also that older citizens remain productive in the workplace.

Workforce demographics are discussed widely, partly because of the anxiety organizations suffer due to the fear that digital ignorance will leave them disconnected from the current teenagers and pre-teens, who form not only their potential workforce but their future marketplace as well. This concern spurred Barclays on to sponsor an 'Open Innovation' challenge on the OpenIDEO platform with the Work Foundation and IDEO, around the theme 'How can we equip young people with the skills, information and opportunities to succeed in the world of work?' The platform crowdsources responses, all conducted online.[19]

This produced six winning approaches and Barclays took one of these concepts, the YouthCafe, and turned it into the Ideas Café, a UK event held simultaneously in London, Coventry, Newcastle and Cheshire, for 500 Barclays apprentices. The event was hosted by Barclays Retail Chief Executive, Ashok Vaswani, in 2013, and was themed around 'The Big Apprentice Conversation' about how apprentices might connect with and inspire friends and communities.[20] The bank has also created 3,500 'Digital Eagles' within branches, young employees who act as social media and digital champions to help inexperienced customers, particularly older ones, use Barclays' digital banking products as well as other online-related advice such as social media. The aim is to provide personalized banking that improves the customer experience and drives retention.[21]

There is no economic 'silver bullet' in expecting that every teenager becomes a Nick D'Aloisio or Max Bryant, but there is a shift of power down the ages happening. Not only were these stories both only made possible by the Digital Renaissance, but also the ability for school-age children to create social networks, publicize themselves and gain a stake in the older economic world is something we have never seen before. In the digital world, age becomes less relevant because content and communication flow more easily than in the physical world. The pattern of work stretching is a sign of how work itself is being recast in the Digital Renaissance.

Freelancing the Organization

Another trend is the freelancing of work as many of those in employment shape a work style more akin to that of a freelancer, flexing their schedule and projects to suit their abilities, lifestyle and preferences. A typical example is Dianne Wentworth, Digital Workplace Manager at AT&T, who has been working remotely for 13 years, based in Dallas with a home office in the garden of her home, working on projects and with teams around the world at hours that suit both her and AT&T. This pattern of restructuring work to suit both corporate and personal needs can also be seen across large companies like McDonalds, which operates an online scheduling system for hourly paid staff and offers a 'passport' that allows young staff to travel and work around the world.

This growth of a more fluid workforce has catapulted agencies like oDesk, Elance and freelancer.com to rapid success. For more than a decade these freelance service marketplaces have been matching freelancers with different skills and services to companies and individuals seeking to outsource work. These sites have grown in size, along with the global pool of freelance talent, sometimes at an exponential rate. For example, oDesk grew by 90 per cent between 2010 and 2012 alone.[22] Elance and oDesk are now merging, creating the opportunity to provide services to larger companies.

At the same time, there is a shift towards insourcing. In one major financial organization, a community was created with an expertise-finder for the insourcing of tasks that might otherwise have been outsourced to a third party at a cost. This is a great example of bringing a more freelance culture inside the organization, enabling employees to utilize their full set of talents and interests in their work. In fact, companies may well already have some of the smartest people working for them but be tapping into only a fraction of their potential; the inspiring effect of the digital work ethic can liberate this latent talent.

While most companies using freelance marketplace services tend to be small, increasingly large enterprises, such as The Motley Fool, are now finding advantages in freelance labour. The Motley Fool is a financial services multimedia company that delivers advice, content, online products and investment services to global investors through a number of different branded websites. To ensure a good supply of content, which helps drive traffic to the sites and ultimately maintains the Motley Fool brand, it runs a blog network using a number of freelance paid contributors. Running this network has been very time-intensive, due to, for example, the need to maintain contracts, manage relationships, oversee editorial processes and issue payments, and these problems have intensified as The Motley Fool continues to scale up its activities. As a solution, The Motley Fool partnered with Elance to create a secure online platform to aid communication, distribute work, produce data and pay the global community of freelance 'talent'. The platform also gives visibility around who is doing what work and keeping track of relative payments. Overall, the system is far more efficient and flexible, and enables scalability. It has also resulted in The Motley Fool leveraging other skills and areas of expertise that bloggers have, as well as the ability to search for specialists in particular territories.[23] The WordPress company, Automattic, is another powerful example of how the new flexibility of work, time and location is creating different models for both employees and external providers, where the lines between these two historically separate groups are blurring.

Freedom from the Working Week

One question that intrigues me is 'How much work is enough work?' If you are unemployed, would five hours of work a week mean you were now part-time or still virtually unemployed? It depends on rates of pay naturally, but it is also a matter of perception. Unemployment feels demeaning and work plays a key role in driving our sense of value and worth. So, how does our quality of life change as we work 10 hours,

15 hours, 21 hours, 30 hours, 60 or 80 hours a week? How far your work covers your economic needs is a key factor, but in a digital workplace where outputs (what we actually produce in tangible results) are what matter, rather than inputs (sitting at your office desk looking busy), pay can be linked to output rather than time spent. Being busy does not pay the bills, but being productive will.

Andrew McAfee and Erik Brynjolfsson are correct; machines will continue to replace us in work but they will also create new industries and formats of work, and the transition will be gradual. The robots will have new roles; they will not only work with us and support us but these machines will perform tasks we currently call 'our work' as well. So, the question of how we allocate work to citizens becomes a central issue in the Digital Renaissance. For example, we currently have huge numbers of working people putting in 70-hour weeks, with all the stress and burnout that brings, while, on the other hand, legions of young people are without any work at all. This imbalance must be changed for the benefit of both groups and also for social and economic cohesion.

The question of time and the duration of what we call the 'working week' must also be refashioned. The traditional work format of the 9 to 5, 40-hour, five-day working week is being reformed as part of the Digital Renaissance, and the 24/7, 365 days in a year we have available opens up new opportunities for when and how we work. The digital world of work stimulates new options for those willing to respond to the transformations happening, while those who resist will find themselves sidelined. As individuals, so long as we have skills that can command a market, we can offer 'units' of commercial value. However, we can also be viewed in the same light as a table in a restaurant that remains an expensive overhead so long as the restaurant doors are closed. If we open up our time and resources to the global marketplace that never closes, we no longer limit ourselves to applying for the few available jobs worth doing that happen to be physically nearby; we can instead connect ourselves with the vastly greater amounts of viable and sustainable work available globally.

The New Economics Foundation (NEF) has called for a rewriting of the 40-hour week, stating that this format met the needs of the industrial factory and the office but not the current workplace. NEF initially advocated a 21-hour week and an accompanying economic rebalancing, so that 21 hours would not simply pay 21/40 of the full week but a higher proportion through Government subsidy and incentives.[24] NEF now seems to have switched to advocating a 30-hour week, which would reduce the normal working week by 25 per cent, meaning a profound change for everyone, since not only would work for more people be created but it would also allow time for life outside work for family, friends, health and social contribution.[25] But our habits and policies need to change in order for this to work. France's experiment in creating a maximum 35-hour week was ultimately repealed;[26] however, a tolerant and flexible working environment in the UK appears to have turned the country into something of a 'new workplace lab', where innovative practices are emerging to meet economic and social needs. While we have no fewer fundamental challenges than other developed countries, experimentation in these turbulent workplace times helps to prove or disprove the value of new models for work. As I will cover in Chapter 8 on education, it may be time to reject Industrial Age thinking and concepts about work, and to redesign how we both learn and approach work with a Digital Renaissance mentality. Repeating what has worked to date will not guarantee our working future.

Notes

1   The Guardian (29 May 2014) Maya Angelou quotes: 15 of the best. *The Guardian*: http://www.the-guardian.com/books/2014/may/28/maya-angelou-in-fifteen-quotes [accessed 30.06.14].
2   Brynjolfsson, Erik and McAfee, Andrew (2014) *The Second Machine Age: Work, progress, and prosperity in a time of brilliance*. New York: WW Norton & Co.
3   Frey, Carl Benedikt and Osborne, Michael A (17 September 2013) The Future of Employment: How susceptible are jobs to computerisation? University of Oxford: http://www.oxfordmartin.ox.ac.uk/downloads/academic/The_Future_of_Employment.pdf [accessed 29.03.14].
4   The Economist (18 January 2014) The future of jobs: The onrushing rave. *The Economist* Briefing: http://www.economist.com/news/briefing/21594264-previous-technological-innovation-has-always-delivered-more-long-run-employment-not-less [accessed 29.03.14].
5   Kirkland, Rik (May 2013) Connecting Everything: A conversation with Cisco's Padmasree Warrior, Interview by Rik Kirkland. McKinsey & Co Insights & Publications: http://www.mckinsey.com/insights/high_tech_telecoms_internet/connecting_everything_a_conversation_with_ciscos_padmasree_warrior [accessed 29.03.14].
6   Brynjolfsson, Erik and McAfee, Andrew (2011) *Race Against the Machine*. www.RaceAgainstTheMachine.com [accessed 29.03.14].
7   As 2 above.
8   As 4 above.
9   Lardinois, Frederic (7 August 2012) Google's self-driving cars complete 300K miles without accident, deemed ready for commuting. TechCrunch: http://techcrunch.com/2012/08/07/google-cars-300000-miles-without-accident [accessed 29.03.14].
10   Handy, Charles (1995) *The Empty Raincoat: Making sense of the future*. London: Random House.
11   UK Federation of Small Businesses (October 2013) Small Business Statistics. http://www.fsb.org.uk/stats [accessed 29.03.14].
12   Williams, Rob (19 February 2013) More than 1,700 people apply for just EIGHT jobs at Costa Coffee shop. *The Independent*: http://www.independent.co.uk/news/uk/home-news/more-than-1700-people-apply-for-just-eight-jobs-at-costa-coffee-shop-8501329.html [accessed 29.03.14].
13   As 11 above.
14   See ITAMCO case study in: Bynghall, Steve (2014) Success with Enterprise Mobile: How tools for frontline employees drive value. Digital Workplace Group: http://www.digitalworkplacegroup.com/resources/download-reports/success-with-enterprise-mobile [accessed 29.03.14].
15   BBC News (25 March 2013) Yahoo spends 'millions' on UK teen Nick D'Aloisio's Summly app. BBC News Technology online: http://www.bbc.co.uk/news/technology-21924243 [accessed 29.03.14].
16   Walker, Daniela (1 October 2013) A 12-year-old's take on everyday marketing concepts. PSFK: http://www.psfk.com/2013/10/12-year-old-marketing-tips.html#!z8ZbX [accessed 29.03.14].
17   Gye, Hugo (10 November 2011) A well-earned rest! Britain's oldest shop worker retires aged 96 after holding down a job continuously since the age of 14. *Daily Mail* online: http://www.dailymail.co.uk/news/article-2059411/Syd-Prior-retire-Britains-oldest-shop-worker-retires-aged-96.html [accessed 29.03.14].

18   Hawkes, Steve (18 December 2013) Britain's 'social revolution' helps employment climb to a record level of 30 million. *The Telegraph*: http://www.telegraph.co.uk/news/politics/10526036/Britains-social-revolution-helps-employment-climb-to-a-record-level-of-30-million.html [accessed 29.03.14].

19   OpenIDEO website (undated) How can we equip people with the skills, information and opportunities to succeed in the world of work? OpenIDEO: http://www.openideo.com/open/youth-employment/realisation [accessed 29.03.14].

20   OpenIDEO website (undated) Ideas Café connects youth to succeed in the world of work. OpenIDEO: http://www.openideo.com/open/youth-employment/realisation/ideas-cafe-connects-youth-to-succeed-in-the-world-of-work [accessed 19.03.14].

21   Fakhri, Daoud (26 September 2013) Barclays empowers branch staff to boost loyalty. Datamonitor Financial: http://www.datamonitorfinancial.com/barclays-empowers-branch-staff-to-boost-loyalty [accessed 29.03.14].

22   Rao, Leena (9 November 2011) oDesk: Online work market will grow to $1billion by 2012. TechCrunch: http://techcrunch.com/2011/11/09/odesk-online-work-market-will-grow-to-1-billion-by-2012 [accessed 29.03.14].

23   Elance website (2013) How The Motley Fool managers its extended workforce. Elance case study: https://www.elance.com/q/sites/default/files/docs/ptc/The-Motley-Fool-Case-Study-Elance.pdf [accessed 29.03.14].

24   Coote, Anna and Franklin, Jane (13 February 2010) 21 Hours. New Economics Foundation Publications: http://www.neweconomics.org/publications/entry/21-hours [accessed 29.03.14].

25   Coote, Anna (12 March 2014) Reducing the week to 30 hours. New Economics Foundation blog: http://www.neweconomics.org/blog/entry/reduce-the-working-week-to-30-hours [accessed 29.03.14].

26   Samuel, Henry (24 July 2008) France drops the 35-hour working week. *The Telegraph*: http://www.telegraph.co.uk/news/worldnews/europe/france/2453277/France-drops-the-35-hour-working-week.html [accessed 29.03.14].

Chapter 6

Leaders Need Followers

> *When the best leader's work is done the people say, 'We did it ourselves.'*
> *Philosopher and Poet, Lao Tzu*[1]

In the 1980s, management theorists popularized a new term 'Management By Walking Around' (MBWA), which was a pithy way of describing what the likes of business guru Tom Peters considered to be one of the arts of leadership – strolling through the offices, warehouses and factories and chatting to staff. Those were the days when the simple act of walking and meeting employees was enough to build culture, productivity and engagement. Iconic CEOs of the time, such as John Akers of IBM, could in that era demonstrate their accessibility and connection to the day-to-day workings of their organizations purely by leaving the executive floors and touring the IBM locations in the US and worldwide.

The modern CEO and the leadership cadre of large organizations can only look back with nostalgia to such a time when the ideal methods of leading were so straightforward. Today, the C-suite has two large and unique challenges that have never before been faced by leaders (particularly CEOs). Not only is there a physical workplace to lead (as there always has been) but now there is a digital workplace that requires continual attention as well. The workforce is working within both physical and virtual environments and, therefore, even MBWA leadership now means 'walking around' in the digital as well as the physical world.

And if that wasn't bad enough, the second challenge all leaders face today is one that stems from the evolution of the digital world of work: Who are you leading now? When Jack Welch became the youngest CEO ever to lead GE in 1981, at least he knew who he was in charge of – the employees. In 2014, Jeff Immelt, the current CEO of GE, has multiple stakeholders, all of whom require his attention. In a modern organization such as GE, there are still the employees, but then there are also long-term contractors (who can outnumber the staff); a fluid freelance community; and the supply chain (critical to business success) – to name but a few. For a company such as Unilever, this wider audience includes 100,000 people in the supply chain; partners of all types who might manage outsourced logistics; customers, ranging from large entities to consumers, depending on the sector; and the broader 'marketplace'. Viewed through this lens, the employees now look like the most straightforward group since at least they are working for the organization on the payroll.

The reason why this proliferation of people requiring leadership has burgeoned so much is because the digital workplace is blurring the divide between what we mean by 'inside' and 'outside'. Leaders must be present and active in the digital world of work and, in so doing, find that their remit has expanded exponentially. Suddenly, MBWA is not only much harder physically, given globalization and workplace fragmentation, but has also to a large extent been superseded by the essential need for digital leadership. Most CEOs do not yet fully realize the extent of this requirement and the opportunity it offers.

What Does Digital Leadership Look Like?

While there is a distinction between tactical social media use and a sustained, strategic approach to 'digital leadership', it is alarming that, according to a 2013 study by CEO. com, less than a third of CEOs of the 500 highest revenue companies in the US have even one account on a social network such as Twitter, LinkedIn or Facebook – and 68 per cent have no social media presence at all. Just as alarming is that only 5 per cent of CEOs have a Twitter account, despite almost 20 per cent of US adults having one – so your staff, suppliers and customers are there, but you are absent.[2]

The exceptions demonstrate what can be achieved. Richard Branson, CEO of Virgin, has (at the time of writing) just under 4 million followers on Twitter.[3] The brand value to Virgin of Branson's reputation and communication capacity is huge. Even Marissa Mayer at Yahoo!, who tends to divide opinion due to her 'come back to the office' mantra, has over half a million Twitter followers[4] and maintains a Tumblr account. The power of Twitter, LinkedIn or Facebook for such C-suite executives is that they can instantly, and with minimal use of their time, connect with large numbers of people they lead across all sections of their universe. As one *Economist* article in January 2013 put it ahead of that year's Davos gathering: 'How can you be a leader if you do not have followers?'[5]

Digital Leaders Feel Accessible

The truth is that, while the turmoil created by the digital workplace is extreme, it offers great opportunities for CEOs and other leaders to communicate, engage and connect with employees instantly and continually on a global basis. Take Andrew Liveris,

Chairman and CEO at Dow Chemical, a global specialty chemical company. Liveris has regularly delivered an employee blog called 'Access Andrew' to the firm's 54,000 global employees since 2007.

Posts are usually delivered weekly, and the blog receives anything between 15,000 and 25,000 visits per blog post and can receive up to 50 comments, indicating that the blog is popular with employees.[6] One of the most important elements of the blog is that it is authored by Liveris himself so is in his authentic voice, although it is checked by a legal and communications team before publishing. This helps create a sense of connection with a CEO that is unusual in a company of Dow's size and type.

Another critical feature is that employees are encouraged to leave comments. These are moderated, but hardly any have ever been removed. Subsequently, the blog has emerged as a channel for dialogue, with employees regularly asking questions and other leaders contributing answers in the thread.[7, 8] Another important approach is that the blog addresses multiple issues and does not shy away from difficult subjects, for example, executives who have left Dow.

It also covers business issues and gives Liveris' personal opinions, for example, on business books he has read. The blog has helped to revolutionize internal communications at Dow through a direct, open and honest channel that means employees can get to know their CEO a little better and vice versa.

The Celebrity CEO: Marc Benioff

Another more widely known example is Marc Benioff, CEO of Salesforce, who is as much a celebrity in his own right as he is a business leader (akin to Richard Branson). Benioff's high profile and reputation has helped to propel Salesforce.com to its success.

There are several reasons why Benioff is so good at what he does. Many of Benioff's communications are personally delivered and built on his own experiences, so that the history of Salesforce.com is also about him. Like Steve Jobs and Apple, the development and brand of the company is wrapped up with that of the CEO. Some of this has been captured in his book *Behind the Cloud*, which tells the story of how he founded the company in 1999.[9]

Benioff also gives the impression that he is accessible, often communicating via his Twitter channel and regularly interacting with other tweeters as well as posting photos.

A similar approach is taken on his Google Plus page. He often posts messages stating that the best way to get hold of him is via his email address, CEO@Salesforce.com.

The personal touch also extends to his regular keynote at Dreamforce, Salesforce's enormous annual conference for customers, which often involves him leaving the stage and walking through the audience.[10] But perhaps the most significant factor is that Benioff is just a very good communicator and is passionate about what he does. For example, he can deliver a 2.5-hour-long Dreamforce keynote speech without notes. Benioff fully recognizes the importance of communication and once wrote: 'I've always thought that the biggest secret of Salesforce.com is how we've achieved a high level of organizational alignment and communication while growing at breakneck speeds.'[11]

Building Trust Among the Digital Leaders

Trust ratings for the likes of Benioff, Branson and Liveris, based on the annual 'Trust Barometer' published by public relations firm Edelman, will be substantially higher than for the average digitally distant CEO. According to Edelman, only 20 per cent of people surveyed trusted business leaders to tell the truth, just 7 per cent higher than for the constantly derided government leaders; but regular, self-generated communication from CEOs can build trust and at least the appearance of accessibility.[12]

The digital workplace ought to be a blessing. Why live out of a suitcase, continually travelling the globe, in order to meet a fraction of your workforce and a tiny number of key stakeholders, when as CEO you can spend a day a week working from home with your family, connecting with hundreds of thousands of people across all your audiences with greater power and impact, and generating trust through every key stroke?

In DWG, as the CEO, I appreciate the ability to communicate easily with the 70 or so people in our workforce over several continents, while the next minute engaging with 500 customers in 100 client organizations, before tweeting to several thousand via tweets and retweets. The potential scope and power of communication possible in as little as ten minutes is exhilarating – and it works for everything from the more personal or poignant digital conversations to bold and broad announcements.

As mentioned earlier, according to predictions from the highly respected World Future Society, 2 billion existing jobs will disappear by 2030[13] due to such innovations as driverless cars, 3D printing and robotics, with the hope that these will be replaced by new jobs but, more likely, by new forms of work and income as the freelancing of the digital economy evolves. Leadership in this entirely new paradigm, in which CEOs must navigate the transformed digital and physical worlds of work, must evolve. If the likes of

Benioff and Liveris are examples, it is not that the need for leadership is disappearing but that best practices for effective leadership require new digital communication skills. Far more personal openness and a mentality that absorbs each digital workplace innovation with gratitude are essential, rather than a philosophy based on fear and avoidance of the unfamiliar.

This revolution in how leadership is practised is being accelerated by the new organizational designs we are seeing in technology firms of all sizes. In Amazon, for instance, hierarchy and leadership exist but the power structure is thin and transparent. This form of leadership is described in new management theory as 'Servant Leadership', where the aim of senior leaders is to support those within the organization, who in turn focus on improving their own tiny part of the service. In the case of Amazon this might be working on the optimum position of a screen button or evaluating the best 'bottom of screen' deals.[14] New structures require new leadership formats.

So, if you are a young manager rising up inside GE, Procter & Gamble or Amazon today, where should you look to learn about the methods of effective leadership in the digital age? You could look to a CEO with minimal presence on Twitter and other social channels … or you could, more usefully, turn to examples from the media and popular culture. You do not find Marc Benioff and Richard Branson taking lessons from the 'leaders without followers' seen at Davos, but instead from singers like Beyoncé who use social media to communicate with millions. These leaders know that, given the economic transformation we are experiencing, entirely new communication styles and patterns for leadership are vital.

Notes

1   Lao Tzu (undated) *The Way of Lao-Tzu*. Pearson, published 1963.
2   CEO.com (2013) Social CEO report: Are America's top CEOs getting more social? CEO.com: http://www.ceo.com/social-ceo-report-2013 [accessed 29.03.14].
3   Richard Branson Twitter account: https://twitter.com/richardbranson [accessed 29.03.14].
4   Marissa Mayer Twitter account: https://twitter.com/marissamayer [accessed 29.03.14].
5   The Economist (21 January 2013) The World Economic Forum in Davos: Leaders without followers. *The Economist Notebook*: http://www.economist.com/blogs/newsbook/2013/01/world-economic-forum-davos [accessed 29.03.14].
6   Sebastian, Michael (25 May 2009) CEO blogs: What you should know. Ragan's HR Communication: http://www.hrcommunication.com/Main/Articles/CEO_blogs_What_you_should_know_2219.aspx# [accessed 29.03.14].
7   Dreher, Joyce (21 January 2013) Linda Kingman on Dow Chemicals' internal blog: Access Andrew. The power of communication: http://4-rf-am.info/linda-kingman-on-dow-chemicals-internal-blog-access-andrew [accessed 29.03.14].
8   The Dow Chemical Company (undated) Access Andrew overview. Dow Chemical Company's Access Andrew blog: http://dialogue.golinharris.com/misc/womma/dow/Access_Andrew_Overview.html [accessed 29.03.14].

9   Benioff, Marc and Adler, Carlyle (2009) *Behind the Cloud: The untold story of how salesforce.com went from idea to billion-dollar company – and revolutionized an industry*. San Francisco: Jossey-Bass.
10   Gallo, Carmine (19 September 2012) Marc Benioff wows his audience by leaving the stage. *Forbes*: http://www.forbes.com/sites/carminegallo/2012/09/19/marc-benioff-wows-his-audience-by-leaving-the-stage [accessed 29.03.14].
11   Benioff, Marc (9 April 2013) How to create alignment within your company in order to succeed. Salesforce blog: http://blogs.salesforce.com/company/2013/04/how-to-create-alignment-within-your-company.html [accessed 19.03.2014].
12   Edelman (2014) Edelman Trust Barometer report: http://www.edelman.com/insights/intellectual-property/2014-edelman-trust-barometer [accessed 29.03.14].
13   Frey, Thomas (3 February 2012) 2 billion jobs to disappear by 2030. World Future Society, Thomas Frey's blog: http://www.wfs.org/content/2-billion-jobs-disappear-2030 [accessed 29.03.14].
14   Deutschman, Alan (August 2004) Inside the mind of Jeff Bezos. Fast Company: http://www.fastcompany.com/50106/inside-mind-jeff-bezos [accessed 29.03.14].

Chapter 7

The Price We Pay

I've occasionally been wrong about certain things, which is in a way more delightful than being right.

Author and digital pioneer, Jaron Lanier[1]

So where's the catch? A Digital Renaissance that makes work fundamentally more rewarding and a wave of innovation as significant as the Industrial Revolution – there must be downsides. Just as the Industrial Age forced an agrarian Britain, through mechanization, into a city-centred landscape, uprooting lives and familiar patterns of living along the way, so the current seismic-level change naturally comes with a 'price tag'.

The 'Second Machine Age', as Brynjolfsson and McAfee call this digital revolution,[2] brings with it great benefits but also costs. In *The Digital Workplace: How technology is liberating work*, I wrote about the challenges of workplace isolation that can be caused by the 'work anywhere' capabilities of an advanced digital workplace. IBM became known as 'I'm By Myself' after a decade of 'home office working'.[3] I also wrote about the addiction we are seeing due to the 'work anytime' options the digital world of work enables. In the two years since that book was published, isolation and addiction have come to be viewed as two core consequences of the digital age – part of the emerging range of 'prices we pay' for the liberation the digital workplace brings.

Alone Again

The fragmentation that may be experienced when we see those we work with less and less in person can be a challenge in any company. There are colleagues in DWG (and I am one of these myself at times) who can find themselves working from home for days at a time and sometimes this can result in a lonely working day, depending on family movements and colleague interactions. Yes, the office each day, every day was a drag, but working in homes not originally designed as workplaces can also bring a sense of detachment from other human beings. To counter this in DWG, we schedule as many in-person days, meetings and weeks as we can. We get together in conjunction with client projects, at member meetings and other ad hoc events and meetings. As mentioned earlier, as these new ways of working bed in, we will need an increased range of viable options for where to work aside from our homes or a commute to a partially empty office building. In the fragmented digital workplace we need to focus on face-time, to make better use of improved real-time audio and video, and to realize that isolation is an issue requiring attention in order to guard against its harmful effects.

As a species we are currently struggling to understand and design our relationship with technology during the digital revolution we are living through. Admittedly we have experienced apparent threats from innovation before (from the mechanization of agriculture to the Industrial Revolution) and some of these fears were justified (for example, the corrosive effects on the human experience of the factory production line). Others turned out to be exaggerated (such as fears that the arrival of the PC would reduce employment levels, whereas in fact employment rose through increased demand and consumerization). However, the ability for technology to seep into every aspect of our lives now, is without precedent. Research suggests that one in 12 minutes of each of our waking hours is spent online in some form or another,[4] so it is no wonder our relationship with technology is a defining issue of our age.

The Paradox of Passion

There is a paradox in the Digital Renaissance of Work. As this new work ethic extends its reach, our personal relationship with work improves. Like artists and artisans, we then develop an intensified desire to work more, showing increased passion for our work. However, this new connection with our work can breed complications. The artisan derives levels of satisfaction from their endeavour, working with greater intensity and duration. The paradox is that the more we enjoy our work, the more we choose to work and this creates the potential for the damaging effects that can stem from work addition.

While the digital workplace may enable the flexibility and autonomy most of us desire, this easy access to work is due to the removal of some of the externally imposed boundaries such as physical location and the commute to a defined workplace – which, ironically, can lead to a new set of problems. Without such constraints, we are left to our own choices – always a dangerous situation. At school we received little or no guidance about managing work–life balance, or the blurring power of workplace technology, so we develop our own habits as we go. And some of these habits are damaging.

We have decided to work more hours, often justifying this action by telling ourselves this is of our own choosing. The simultaneous blessing and curse of the digital world of work is the permanent availability of its tools, and this lethal cocktail has eroded evenings, weekends and vacations as work increasingly pollutes our 'off time'. These new patterns of work, particularly for those in their 30s and older, have produced unprecedented levels of work addiction and a marked deterioration in our ability to distinguish between work time and non-work time. The older ages I refer to have had no schooling or training in navigating this new digital workplace capability, but there are some encouraging signs from teenagers and those in their early 20s, representing the second wave of digital natives, who regard switching off as a conscious act in a desire to gain control over the digital world, rather than be at its mercy.

A Symptom of Our Lifestyle: The Rise of Digital Addiction Clinics

Education, training and a new set of mature digital work habits are required for the future digital workforce (and the companies they will work for) to produce a more sustainable balance between work and living. Too many work addicts tell me they are making a choice rather than being compelled to work – but that is what addicts do, they protect their addictive habits. Certainly, there is not yet a consensus about the best way to treat this new digital addiction. Is 'connection addiction' a clinical condition or just a habit that can be changed more easily? Persistent digital connection is already viewed as an addiction in countries like South Korea, where dedicated facilities are provided, but the condition is not yet formally recognized by the American Psychiatric Association.[5]

Despite this, dedicated programmes and centres tailored towards treating digital addiction are emerging in the US. One example is the 'Internet Addiction Program' at Bradford Regional Medical Center, Pennsylvania, a ten-day programme involving a 72-hour detox and various therapies.[6, 7] The programme's director, Dr Kimberly Young, believes the Internet is a 'new outlet for traditional addictions'.[8] Meanwhile, the reSTART centre, based in Washington State, encourages you to 'disconnect and find yourself' for 'a sustainable lifestyle' and offers a residential retreat with counselling, coaching, 12-step programmes and other activities. There is also a range of other assessment, mentoring and support options, including meetings of Internet & Tech Addiction Anonymous (ITAA). Ironically, you can apply for these programmes online![9]

Another high-profile expert in Internet addiction is Dr Dave Greenfield, author of *Virtual Addiction*.[10] He offers a range of consultations and therapies, including two- or five-day intensive treatment programmes – one for general technology addiction and another for video game, computer and Internet addiction – to help individuals 'plug back into life'.[11]

Enough is Enough: Volkswagen Turns Off Email after Hours

One of the most obvious examples of the blurring of work and personal lives is through the checking of email and other digital workplace services on mobile devices outside working hours. This activity has become so normalized that a 2013 survey suggested that 54 per cent of US adults had bosses who actually expected them to work during vacations, despite the fact that 67 per cent acknowledged that other family members get upset when they check their emails at times like this.[12]

At an organizational level there are very few cases in the public domain of anything being done about working out of hours – so far. One notable exception is Volkswagen in Germany, which, in 2011, agreed to turn off its Blackberry servers to some workers so that they could only receive email up to half an hour before and after their shifts. This change is an official agreement with the Workers Council at Volkswagen and excludes some work groups such as senior management.[13] Workers Councils in Germany yield considerable influence on working conditions and practices within organizations.

Many observers have warned that Volkswagen's approach may not work for other organizations and could actually be unpopular with workers who value the flexibility of 24/7 access. A slightly different approach has been taken by Deutsche Telekom, which offers a 'Smart Device Policy' stating that workers should declare 'communication free' times outside working hours, with management agreeing not to expect them to answer emails or calls.[14] This approach from Deutsche Telekom feels like a more mature stance to take, with the debate being centred on whatever kind of work pattern suits a specific business unit or team, and able to be varied based on individual requirements. Those with young families or ageing parents, or new single hires, will have a range of ideal designs for their working day and week, and the more these differing preferences can be accommodated, the happier and more productive the workforce will be.

When the Machines Evolve, What is Left for Us?

Another price to pay is encapsulated in this quote from *The Second Machine Age*: 'There is no economic law that says that when technology advances, everybody necessarily benefits: some people, even a majority of people, could be worse off'.[15] IBM has invested US$1 billion in its hyper-intelligent 'Jeopardy'-winning machine 'Watson',[16] while the annual Consumer Electronics Show in Las Vegas is awash with medical robotics that begin to make a visit to your doctor's surgery seem a far less reliable approach than 24/7 self-monitoring technology that can assess and keep you posted on your medical condition – with 100 per cent accuracy and irrespective of surgery opening hours.[17]

The rise of robotics and machine capability over the past 40 years has generally so far been complementary to our human endeavour; for example, in motor manufacturing, the introduction of heavy lifting and assembly gear made work easier for factory-floor workers; while in call centres, technology has allowed staff to retrieve customer data more quickly and accurately, and therefore provide a better service for callers. But digital innovation increasingly means that machines can actually substitute for humans and the effects of this are often beneficial for the customer. The success of mobile banking via smartphones is such that, by the end of 2014, 90 per cent of the US population will even be able to make a deposit via their mobile (by taking a picture of their pay stub or credit card balance).[18] These 'mobile only customers' have no need for a bank branch and a wave of branch closures is predicted. As retail banks provide technology for branch staff to set up visiting customers with mobile phone banking, jobs become redundant because, once the customer is set up to bank via mobile, they hardly ever need to visit a branch.

The reality is that the machines are faster and often better. Why call a busy call centre, listen to the litany of options, eventually reach a human being, navigate all the security protocols, engage in conversation (all requiring our discretionary effort in these time-poor days), just in order to transfer a balance or pay a bill when we could carry out the same tasks in 10 per cent of the time as we sit sipping a cappuccino, before processing ten other similar-level daily living tasks in 15 convenient minutes?

We are starting to see this everywhere. Take digital services that can write well and accurately for 24/7 media, including creating the headlines, which are replacing some (maybe many) journalists. Since as far back as 2010, US-based StatSheet has been producing completely automatically written sports news feeds for different teams.[19] There has been a huge rise in the popularity of so-called 'quantified self' technologies that screen our bodies for early signs of danger across hundreds of key blood, heart and brain indicators, relaying information to us and even to our general practitioner when needed. Why go for a breast screening, smear test, asthma check, cholesterol measurement every two years, or even more frequently, when all this can be done continuously in the background by the latest smartphone apps?

In fact, the acceleration of this kind of activity has caused enough concern around data privacy for the Food and Drug Administration (FDA) in the US to ban genetics company 23andme from providing personal genetic testing to consumers.[20] Nurses, doctors, operatives and administrators will not vanish, but there will be fewer needed.

The self-driving car was first heralded by Google in 2010 as a pivotal technology change and met with deep scepticism; but, four years on, all motor manufacturers know it works – and not just as well as human drivers but better since it removes the big issue of human error that stalks every vulnerable cyclist and the soft skin and bones of pedestrians as we hurtle past at 30 mph and more in our steel boxes. The challenge to the adoption of self-driving vehicles is not a technical one but is instead centred around habit and our perceptions of what feels 'safe'. In fact, early research suggests that driverless cars may be better at keeping a safe distance and more able to react sharply to sudden braking, making these machines arguably safer than their human counterparts.[21]

A Positive Digital Trajectory

Alarming as these stories may be, the Digital Renaissance trajectory remains, in my view, overwhelmingly positive. Interesting developments range from the profound and life-saving, such as the example of 15-year-old Jack Andraka who, with no medical training and using online research only, formulated an early detection system for pancreatic cancer, one of the most lethal cancers, killing 40,000 people each year. Cure of this cancer is particularly challenging because often it is only detected late once the disease has already spread, and so Jack's breakthrough may save hundreds of thousands of lives in coming years.[22] At the more mundane end of the scale, but still incredibly useful, is an app that tells you when the next bus is coming in real-time so you can stay warm and dry at home for longer!

But, as we have seen, there is a price to pay in the loss of established jobs that are evaporating by the day. Just as the Industrial Revolution brought soot-filled air to London and awful levels of child labour exploitation, this revolution comes with some costs. The replacement of people with machines is not a reason to destroy the machines, but rather to see where the pain falls and to adapt our education, organizations, government institutions and policies to tackle the damaging effects. Not everyone can become a hyper-prized Google engineer on US$3 million a year (there aren't many of these, although they do exist), but we can attempt to understand the underlying shifts and to use our uncanny ability to change in order to ensure that we will have lives worth living for the 9.6 billion humans who will occupy this planet by 2050.[23]

The education gap is a chasm, as I write about in the next chapter, with a growing mismatch between what we know and what work requires of us. In 2015, there will be 900,000 unfilled IT positions across the European Union, according to a report by the European Commission on the digital economy.[24] At this stage the impact of the Digital Renaissance is evolving and the consequential effects being noticed. It is naive to simply hope that work will be there for each of us but growth in the digital economy and the evolution of new services for a population with a longer life-span, suggest that the bigger picture is more complex than merely 'the robots are taking our jobs'.

Notes

1   Witchalls, Clint (10 February 2010) Interview with Jaron Lanier, *The Independent*: http://www.independent.co.uk/life-style/gadgets-and-tech/features/jaron-lanier-web-20-is-utterly-pathetic-1894257.html [accessed 01.07.14].

2   Brynjolfsson, Erik and McAfee, Andrew (2014) *The Second Machine Age: Work, progress, and prosperity in a time of brilliance*. New York: WW Norton & Co.

3   Miller, Paul (2012) *The Digital Workplace: How technology is liberating work*. London: Digital Workplace Group: http://digitalworkplacebook.com [accessed 29.03.14].

4   Sweney, Mark (7 October 2013) Britons spend one in 12 waking hours online, pushing ad spend to a record high. *The Guardian*: http://www.theguardian.com/media/2013/oct/07/btitons-online-ad-spemd [accessed 29.03.14].

5   Williamson, Lucy (1 August 2011) South Korean clinic treats web addicts. BBC News Asia Pacific Online: http://www.bbc.co.uk/news/world-asia-pacific-14361420 [accessed 29.03.14].

6   Bradford Regional Medical Center website (2014) Internet Addiction program. http://www.brmc.com/programs-services/internet-addiction-bradford-pa.php [accessed 29.03.14].

7   Neal, Ryan W. (4 September 2013) Digital detox: Pennsylvania opens America's first internet addiction clinic. *International Business Times*: http://www.ibtimes.com/digital-detox-pennsylvania-opens-americas-first-internet-addiction-clinic-1402776 [accessed 29.03.14].

8   Daley, Elizabeth (4 September 2013) Brick-and-mortar clinics treat Internet addicts. Reuters: http://in.reuters.com/article/2013/09/04/usa-addiction-internet-idINDEE9830DV20130904 [accessed 29.03.14].

9   reSTART Center for Digital Technology Sustainability website: http://www.netaddictionrecovery.com [accessed 29.03.14].

10 Greenfield, David (1999) *Virtual Addiction: Help for netheads, cyber freaks and those who love them*. Oakland, CA: New Harbinger Publications.

11 The Center for Internet and Technology Addiction website: http://virtual-addiction.com [accessed 29.03.14].

12 Ricoh (25 June 2013) Who pays the price for 'working vacations'? Ricoh Americas Corporation news release: http://www.ricoh-usa.com/news/news_release.aspx?prid=1068&alnv=pr [accessed 29.03.14].

13 BBC News (23 December 2011) Volkswagen turns off Blackberry email after work hours. BBC News Technology online: http://www.bbc.co.uk/news/technology-16314901 [accessed 29.03.14].

14 Sheahan, Maria (23 December 2011) VW agrees to kick the 'Crackberry' habit. Reuters: http://www.reuters.com/article/2011/12/23/us-volkswagen-blackberry-idUSTRE7BM0H120111223 [accessed 29.03.14].

15 As 2 above.

16 Leske, Nicola (9 January 2014) IBM to invest $1 billion to create new business unit for Watson. Reuters: http://www.reuters.com/article/2014/01/09/us-ibm-watson-idUSBREA0808U20140109 [accessed 29.03.14].

17 Kelion, Leo (9 January 2014) CES 2014: Phones morph into 'stun guns' and 'tricorders'. BBC News Technology online: http://www.bbc.co.uk/news/technology-25663424 [accessed 29.03.14].

18 Groenfeldt, Tom (6 January 2014) Mobile phones set to reduce banking branches in 2014. *Forbes*: http://www.forbes.com/sites/tomgroenfeldt/2014/01/06/mobile-phones-set-to-reduce-banking-branches-in-2014 [accessed 29.03.14].

19 Schonfeld, Erick (12 November 2010) Automated news comes to sports coverage via StatSheet. TechCrunch: http://techcrunch.com/2010/11/12/automated-news-sports-statsheet [accessed 29.03.14].

20 Fielding, Matt (5 December 2013) G23andme genetic testing banned by FDA. Calculator.co.uk: http://www.calculator.co.uk/2013/12/23andme-genetic-testing-banned-by-fda [accessed 29.03.14].

21 Sanghani, Radhika (29 October 2013) Google's driverless cars are 'safer' than human drivers. *The Telegraph*: http://www.telegraph.co.uk/technology/google/10411238/Googles-driverless-cars-are-safer-than-human-drivers.html [accessed 29.03.14].

22 BBC News (21 August 2012) US teen invents advanced cancer test using Google. BBC News Magazine online: http://www.bbc.co.uk/news/magazine-19291258 [accessed 29.03.14].

23 UN News Centre (13 June 2013) World population projected to reach 9.6 billion by 2050 – UN report. *UN Daily News*: http://www.un.org/apps/news/story.asp?NewsID=45165 - .UygzRPl_s34 [accessed 29.03.14].

24 BBC News (5 March 2013) EU Commission launches 'grand coalition' to tackle IT shortage. BBC News Technology online: http://www.bbc.co.uk/news/technology-21668166 [accessed 29.03.14].

Chapter 8

Education – The Revolution Starts Here

The only thing that interferes with my learning is my education.

Albert Einstein[1]

If the world of work is being transformed, the changes starting to sweep through the education system – a structure that has remained little changed at school and university level in more than 200 years – may be even more profound. This is because dramatic shifts in how, when and where the youngest in our societies are educated, will affect the most impressionable globally. The good news is that what happens to education in the coming years has the capacity (for those who seize the opportunity) to make learning for all ages a far more stimulating, accessible and relevant experience.

In terms of education we are currently in a position analogous to the era when the very first 'steam-powered horseless carriages', running on internal combustion engines, appeared back in 1807.[2] At that time, learning to ride a horse was still an essential life skill. Few children went to school but instead learnt the mainly rural skills practised by their parents from their elders. In tune with the profound societal changes industrialization brought with it, learning then adapted to a form more suited to the Industrial Age, providing the skills required for manual labour and crafts, as well as managerial and administrative capabilities. However now, the content, structure and design of today's schools increasingly appear as a relic of a disappearing era. This is perhaps inevitable in that education has always lagged behind technological and business changes. It is not that teachers are failing to do their best to educate students; it is more that there is a growing gulf between the skills our Digital Renaissance world needs and those being taught.

Teachers are Everywhere

As Brian Solis says in the Foreword, the scale of the changes required is profound – and not likely to be welcomed with open arms by the institutions affected. How should business schools react to the experiment in 2012 of Anant Agarwal, Professor of Computer Science at MIT, teaching his module 'Circuits and Electronics' digitally? Typically for MIT, this was a very challenging course and yet more than 7,200 students out of the 155,000 students who enrolled from 162 countries passed the exams – all through a purely online teaching programme. Now this may sound like a low pass-rate from the total students involved, but take a moment to consider the scale of the operation: there were more pass-grade students in that one term than could be taught physically by Agarwal in an entire 40-year teaching career; and more students took part in this one course than all MIT alumni in total – all of this achieved with no marketing budget.[3]

The corporate world (that of the marketplace and 'customer pull') is quite well acclimatized to a natural rhythm of change but education – for all age groups – is not so familiar with this pattern. Governments and curriculum boards tend to set standards that run for years and decades. But what is needed today is accelerated change because the Digital Renaissance of Work clock is ticking. As mentioned before, the World Future Society predicts that more than 2 billion jobs will have disappeared globally by 2030 – along with more than half of today's Fortune 500 companies and over 50 per cent of all higher education establishments.[4] It is easy to throw statistics about but these job trends are also supported by the *The Economist* magazine.[5] The disappearing roles will be partly due to automation but also to the widely reported chasm between the skills required and the available pool of people with the education to match those needs. In the 2013 Manpower Talent Survey, a third of the 38,000 employers surveyed worldwide were experiencing difficulties in filling jobs due to lack of available talent.[6] British inventor James Dyson says 31,000 engineering jobs remain unfilled in the UK in 2014 due to a lack of capable applicants.[7] So, in many countries, we are seeing both chronic long-term unemployment (now affecting young people profoundly in countries such as Spain) alongside a mounting number of unfilled vacancies. What we know and what is needed are out of sync.

Schools and teachers attempt, within the current restricted examination-based format, to engage and stimulate pupils. They use collaborative techniques and narrative-led styles to make learning as enriching as possible. But the education structure and much of the content today, in both the developed and developing worlds, was designed for a bygone age. Perhaps rather over-stated, but Sugata Mitra, a pioneering educationalist in India whose talk at the TED conference attracted 2 million views in 2013, says: 'Schools are not broken, they are just obsolete.'[8]

Designed for a World That Has Vanished

The current school format (and much of higher education) is a product of the British Empire, stretching back 300 years, when a regimented system of teaching was introduced to produce people as parts for the industrial machine. Reading, writing, mathematics and information were required, taught in set blocks of time, to pupils in rows of desks with a teacher in command at the front. And it worked extremely well, educating our young for a plethora of defined jobs and, particularly in the early and middle parts of the twentieth century, our productivity and quality of living increased.

Just as the Industrial Age established formal education for the young, so the Digital Renaissance will in time redesign how we all learn. What is exciting is that this redesign will, in my view, happen across all age groups from the very young to the still (mentally capable) very old. Equally transformational is that, with the advent of digital education, the poorest in our world – who have historically been excluded from high-grade teaching – will have equal access to world's best education. A poor nine year old in Namibia can now, thanks to digital connections, enrol into Khan Academy (as described below) and learn in exactly the same way as a wealthy child of the same age living in the smartest part of Paris. The implications for our planet of the sustained and widespread effects of such 'democratic learning' will be remarkable.

As regards educational content and format, here are my predictions based on what I believe the new digital world of work economies will require.

- We will see rote learning decline and critical thinking skills increase as our children are increasingly educated in ways that seek to differentiate them from the hyper-smart artificial intelligence systems they will work alongside.

- Learning (as happens through online courses) will be more personalized than has been possible with a teacher and 30 children in a classroom.

- The focus will be on early years creativity and capabilities, because so much of how we think evolves when we are small children.

- The concept of 95 per cent of education happening when we are under 21 years of age will morph into a life-long learning experience.

- The issue of the skills (and what is dubbed the 'mindset') gap will be felt ever more acutely within organizations as the younger demographic works easily and collaboratively with evolving technologies, while those who have been in the workforce for 20 or more years struggle to change their habits and to relearn.

Revolutionizing learning: Khan Academy

Probably the most high-profile early stage change in educational format we have seen so far has been through so-called 'Massive Open Online Course' (MOOC) platforms, such as Coursera, Udacity and Future Learn, which are revolutionizing the ways in which people study. These have their critics, who argue that smaller-scale and less open online learning is powerful also, but the current MOOCs host high-quality courses, which are designed by leading universities and open to everybody online, often utilizing video and providing some interaction with tutors. Not only does this provide education for all, but it also changes the way learning material is delivered, with the potential to be highly disruptive for the higher education sector.

Much of the thrust behind the MOOCs has been from not-for-profits. One of the early pioneers was Khan Academy. The Academy's stated mission is 'the goal of changing education for the better by providing a free world-class education for anyone anywhere'.[9] Khan Academy features thousands of learning resources with mainly short videos, interactive quizzes (with badges issued for getting the answers right) and opportunities to chart progress. Material is translated into multiple languages and there is a guarantee that all content is free and always will be.

Salman Khan, a New York-based hedge fund manager at Accenture, founded the Academy in 2008. The idea had grown out of Khan's virtual tutoring of various cousins and family members, usually through videos, which were being posted onto YouTube. In 2009, Khan quit his job to grow Khan Academy and, by 2010, Google had contributed a grant of US$2 million and the Bill and Melissa Gates Foundation US$1.5 million to fund its growth.[10] The latter came about partly because the Gates' own children were already using the Khan Academy as a study aid.[11] Khan still dominates the site and personally narrates many of the video resources.

One of the aims of Khan Academy is to 'flip' the classroom so resources can be watched at home, although they can also be used by teachers within the classroom. While this approach has not enamoured all educationalists, some of whom argue that Khan is reinforcing rote learning, [12] the platform has proved popular. By late 2012, there were 6 million unique visitors accessing Khan Academy each month, contributing to more than 200 million video viewing sessions. It is estimated the material is used in around 20,000 classrooms globally.[13]

Technology is also transforming some schools themselves with encouraging results and creating environments that put the digital workplaces of some larger commercial organizations to shame. For example, the Calgary Board of Education in Canada has created a beautiful-looking SharePoint environment, called Iris, used by both children and teachers to encourage 'personalized' learning.[14]

Individual schools have sought to use technology to help educate pupils and to accelerate learning but very few have done so as comprehensively or as effectively as the Essa Academy. This school, based in Bolton in the north west of England and open to

11 to 16 year olds, is almost certainly the most technically advanced school in the UK. To start with, every teacher was issued with an iPad and every pupil with an iPod Touch. The students can email teachers with questions and keep in touch, and it helps them with their homework. Initial responses from pupils were enthusiastic, with incidents reported such as students playing football while listening to a podcast about atomic structure through headphones as part of their exam revision.[15] After this initial success a decision was made to extend iPads to all pupils.

Screens in classrooms help connect to resources already available on the iPads and staff create interactive textbooks, which include video and other rich media. Moreover, all the schools' courses are available on the iTunes U app, an educational facility that places all the materials in one place. iPads are regularly used in class, for example, to film and play back learning sessions.[16, 17] The introduction of technology has helped contribute to a dramatic rise in performance: the Essa Academy was previously a failing school that had been closed but, in 2012, every pupil achieved five A* to C grade passes at GCSE, compared with a former rate of just 40 per cent. Even more amazingly, the new technology has also contributed to an overall decrease in operating costs at the school.[18]

New Roles for Teachers

More generally though, while adults are experiencing increased flexibility around when, where and how they work, their children are so far enjoying few, if any, of these beneficial changes to their 'working day'. Mitra's studies using a 'self-organized learning environment' (SOLE) have demonstrated the remarkable power of children to learn independently when left with a street-side computer in India. In Mitra's work, exam attainment levels for learning without any teacher apart from the Internet match those of privately educated Mumbai children with dedicated teachers. It seems that children working in small self-organized groups can navigate the web and learn as much as, and sometimes even more than, the regimented education system provides at huge cost and resource. The legendary scientist and writer Arthur C Clarke, echoing Mitra's work, said that 'Any teacher that can be replaced by a machine should be!'[19] The teacher in a SOLE (and similar results to the Indian prototype studies have now been achieved in many countries) becomes someone to encourage and to pose intriguing questions, such as one posed by Mitra, in English, to a group of non-English speaking Italian 14 year olds: 'Who was Pythagoras and what did he know?' Without any supervision, this challenge resulted in detailed, accurate answers – in English – within 30 minutes.

SOLEs are a useful experiment about what happens when powerful, freely available technology is open to children. But it also echoes the discovery by Jake Andraka at age 15 of his method for the early detection of pancreatic cancer, achieved purely via an Internet connection in his bedroom. Or the tales we hear from our own children, who have taught themselves photography or to play the guitar by watching videos on YouTube. The questions for education that technology is surfacing are profound. Why send children to school five days a week, from 8am to 4pm, when their parents work from home on two days a week? Why sit bored, listening to a dull, and maybe not very good, Spanish language teacher in a classroom of other 13 year olds, when one of the most charismatic teachers of that subject could teach you virtually, at a time of your choosing, in a place that you find conducive to learning? Why have a teacher at all when children can work effectively in small self-organized groups and learn via the Internet, so long as they have engaging questions and encouragement?

Parents talk enthusiastically about the positive effects of work–life blurring, so why should their children be deprived of that privilege? Given the evidence of the dramatic mismatch between the outcomes schools are producing for our children and what we need as a society, the case for change is compelling. At a Davos meeting in 2013, Khadija Niazi, a 12-year-old schoolgirl from Pakistan shared a stage with such luminaries as Bill Gates and talked about the free university courses she can view through online platforms, and how she has scored highly in physics. This is a fundamental shift in access to education.[20]

In an Internships.com study of more than 1,345 students, 50 per cent of students answered that they don't need a physical classroom in order to learn; 53 per cent consider online colleges to be just as reputable as traditional colleges; and more than a third see the future of education as becoming more virtual.[21] They do say that most students currently consider a classroom setting as important despite the availability of online routes – but this will inevitably change as technology creates better user experiences to connect students and teachers.

What we are currently seeing is boredom, children refusing school or attending reluctantly, universities unable to prove their courses are worth the fees, MBA courses that add little value, and syllabuses that fail to allow our enthusiasms to flourish. What we need is a revolution in how we learn when we are young, and in how our desires and needs to learn can persist as we grow. The signs of strain and deterioration are widespread.

Alternative approaches are springing up, such as one initiative from Peter Thiel, the billionaire entrepreneur who cofounded PayPal and an early investor in Facebook. Thiel has set up the Thiel Fellowship, which gives out 20 US$100,000 grants each year to young entrepreneurs under the age of 20, to allow them to 'skip college and focus on their work, their research and their self-education'. The selected 20 are mentored with a series of high-profile thinkers and entrepreneurs, who aid with advice and connections.[22]

Thiel has some robust views on the quality and value of higher education. He has been quoted as saying: 'If Harvard were really the best education, if it makes that much of a difference, why not franchise it so more people can attend? Why not create 100 Harvard affiliates? It's something about the scarcity and the status. In education your value depends on other people failing. It's a way to ignore that people are falling through the cracks, because you pretend that if they could just go to Harvard, they'd be fine. Maybe that's not true.'[23]

New School Formats – More Local, Smaller Scale

If students are learning from teachers online, or can take the SOLE approach, without a teacher at all or with the teacher simply acting as an encouraging voice, what will my role be, teachers are wondering? The teacher role will evolve to meet the new educational styles and needs in a way that may well liberate them from administering rote learning, connecting them into a potentially more fulfilling coach-like role.

Schools could (as some already do partly) open into the evening and seven days a week, offering a place where children can choose to come to learn, using the best technology available. With diligence for security and oversight, these new-style learning centres could equally well be open to adults of any age to learn too. We are so paranoid about what might happen if mature, retired people enter a school that we forgo the latent potential for huge accelerations in relevant learning organized together, while also tapping into the central knowledge brain of the Internet.

Not only could how we learn change dramatically but the flexibility in education would beneficially support relationships between children, parents and their ageing parents, as the working week selected out of seven days could be flexed to suit the family needs. This does not amount to a loss of the structure that children need in order to grow up securely and confidently, but instead allows choice, enabling young people to learn with passion rather than through compulsion. Schools (as we have seen with large centralized offices) may fragment to provide smaller, more numerous learning spaces, close to where children live, promoting better relationships in neighbourhoods (and, as we have seen somewhat in the US, with unused shops in high streets being turned into 'micro schools'). A 12 year old may learn better for a few hours a day, enjoying 'going online for learning' in the early afternoon but looking forward also to their physical and social interaction on Monday, Thursday and Friday. And, where couples separate but struggle to share access given the rigidity of the school format, they are able to shape each week to enable both parents to have more time with their children.

The drudgery of term time versus holidays, or working days versus weekends, can be replaced by a more fluid rhythm for the year, more in tune with the seasons and the increasingly volatile weather we all have to live with. The structure and content of our education system must be transformed and recalibrated for the digital age, however the process of this change may prove even more challenging than that sweeping through the working world. The positive implications though could be even more far-reaching as education shifts from necessity and drudgery to freedom, passion and power.

Notes

1   The Telegraph (30 March 2012) Albert Einstein: 10 of his best quotes, *The Telegraph*: http://www. telegraph.co.uk/science/science-news/9176616/Albert-Einstein-10-of-his-best-quotes.html [accessed 29.03.14].

2   François Isaac de Rivaz designed the first car powered by an internal combustion engine in 1807, well before the later petrol-driven automobiles; see Wikipedia: http://en.wikipedia.org/wiki/François_Isaac_de_Rivaz [accessed 29.03.14].

3   Anant, Argawal (June 2013) Why massive open online courses (still) matter. TED talk transcript: http://www.ted.com/talks/anant_agarwal_why_massively_open_online_courses_still_matter/transcript [accessed 29.03.14].

4   Frey, Thomas (3 February 2012) 2 billion jobs to disappear by 2030. World Future Society, Thomas Frey's blog: http://www.wfs.org/content/2-billion-jobs-disappear-2030 [accessed 29.03.14].

5   The Economist (January 2014) The future of jobs: The onrushing rave. *The Economist Briefing*: http://www.economist.com/news/briefing/21594264-previous-technological-innovation-has-always-delivered-more-long-run-employment-not-less [accessed 29.03.14].

6   ManpowerGroup (2013) Talent Shortage Survey 2013 Research Results. ManpowerGroup: http://www.manpowergroup.com/wps/wcm/connect/587d2b45-c47a-4647-a7c1-e7a74f68fb85/2013_Talent_Shortage_Survey_Results_US_high+res.pdf?MOD=AJPERES [accessed 29.03.14].

7   The Engineer (5 February 2014) UK kicking out engineers to work for competitors says Dyson. www.theengineer.co.uk: http://www.theengineer.co.uk/channels/skills-and-careers/news/uk-kicking-out-engineers-to-work-for-competitors-says-dyson/1017968.article [accessed 29.03.14].

8   Mitra, Sugata (February 2013) Build a school in the cloud. TED talk transcript: http://www.ted.com/talks/sugata_mitra_build_a_school_in_the_cloud/transcript [accessed 29.03.14].

9   Khan Academy (2014) A free world-class education for anyone anywhere. About Khan Academy: https://www.khanacademy.org/about [accessed 29.03.14].

10   Khan Academy (undated) How did Khan Academy get started? Khan Academy FAQ: http://khanacademy.desk.com/customer/portal/articles/329316-how-did-khan-academy-get-started [accessed 29.03.14].

11   Thompson, Clive (15 July 2011) How Khan Academy is changing the rules of education. *Wired*: http://www.wired.com/magazine/2011/07/ff_khan/all [accessed 28.03.14].

12   As 11 above.

13   Noer, Michael (2 November 2011) One man, one computer, 10 million students: How Khan Academy is reinventing education. Forbes: http://www.forbes.com/sites/michaelnoer/2012/11/02/one-man-one-computer-10-million-students-how-khan-academy-is-reinventing-education [accessed 29.03.14].

14   Habanero Consulting (undated) Calgary Board of Education. Habanero case study: http://www.habaneroconsulting.com/member-portals/Calgary-Board-of-Education [accessed 29.03.14].

15   Apple (undated) An ailing UK school makes an incredible transformation. Apple in Education profiles: http://www.apple.com/uk/education/profiles/essa [accessed 29.03.14].

16   Cellan-Jones, Rory (12 December 2012) iSchool – can tech really deliver education? BBC News Technology online: http://www.bbc.co.uk/news/technology-20667870 [accessed 29.03.14].

17 Garner, Richard (20 March 2012) The school where every teacher has an iPad … and every pupil has an iPod. *The Independent*: http://www.independent.co.uk/news/education/education-news/the-school-where-every-teacher-has-an-ipad-and-every-student-has-an-ipod-7578167.html [accessed 29.03.14].
18 As 17 above.
19 The Arthur C Clarke Foundation (2014) Sir Arthur's quotations: http://www.clarkefoundation.org/sample-page/sir-arthurs-quotations [accessed 29.03.14].
20 Haq, Riaz (February 2013) 12 year old Pakistani girl shares online education experience at Davos. Haq's Musings: http://www.riazhaq.com/2013/02/12-year-old-pakistani-girl-shares.html [accessed 29.03.14].
21 Schawbel, Dan (11 June 2013) Millennials believe the future of education will be virtual. *Forbes*: http://www.forbes.com/sites/danschawbel/2013/06/11/millennials-believe-the-future-of-education-will-be-virtual [accessed 29.03.14].
22 Thiel Fellowship (undated) About the Fellowship: http://www.thielfellowship.org/become-a-fellow/about-the-program [accessed 29.03.14].
23 Lacy, Sarah (10 April 2011) Peter Thiel: We're in a bubble and it's not the Internet, it's higher education. Techcrunch: http://techcrunch.com/2011/04/10/peter-thiel-were-in-a-bubble-and-its-not-the-internet-its-higher-education [accessed 29.03.14].

Chapter 9

A Future Fit for Work

In the second part of this book, my co-author Elizabeth Marsh will lead you carefully through the key elements of establishing a digital workplace worth working in. Elizabeth offers a roadmap for a Digital Renaissance of Work, not just for you and your colleagues, but for the many others who make up your world of work too.

Creating a human-centred digital workplace is essential for any organization to evolve, flourish and remain successful in the new world of work. But that is only part of what is possible at this pivotal time in human history. As Brian Solis says in the Foreword, we find ourselves working and living at a unique moment where 're-skinning' the past is rapidly proving to be futile.

What I have tried to express so far is that the Digital Renaissance of Work may have been initiated through technological innovation, but is now precipitating and fuelling a complete recasting of work into something that human beings want, need and will thrive upon. Freedom, passion, fulfilment and meaning are fundamental to a life well lived – and now, rather than something we purely associate with life outside of work, these elements can become part and parcel of our daily working experience. To have even some degree of this opportunity percolate into our working hours turns this into a journey worth making.

My eldest daughter, Gabriella, in her first serious job at 21, told me that she would rather gain promotion in her current company than look elsewhere, because she loves both the work and her colleagues. If we can each combine a fair commercial return with work we relish, that makes a huge difference to our experience of being alive. Work still requires effort, struggle and tedium at times (as even hugely successful artists like Damien Hirst admit) but the overall trajectory is one that is producing more and more individual stories that echo Gabriella's. My own experience over 30 years has been a persistent rejection of work that would constrain or degrade me, coupled with a determination to fashion businesses and working cultures for myself and colleagues that would inspire and reward each of us.

The current opportunity in work (and in life) is to engage and appreciate the digital flowering around us. Unlike the first Renaissance, we can use our globally connected planet to observe, examine and enjoy the full impact of the Digital Renaissance we are fortunate to be living through as it unfolds. For organizations of all sizes, irrespective of where they are located physically, the digital transformations required will not only build enduring value and strong commercial returns but will also create a better working experience for those involved – it really is a win–win.

Machines, whether advanced robotics or highly intelligent services and systems, will abound. Sherry Turkle, author of *Alone Together* has written about how we must remain human and connected in a world she feels is becoming *dis*connected, calling for a time of 'stock-taking'.[1]

On a similar theme, my view is that we must now place ourselves as human beings at the centre of the digital worlds that are unfolding, so that the benefits and extraordinary powers of the 'machine world' enhance the quality of being alive as a species. We must work alongside the digital world, while at the same time ensuring that it expands us as human beings rather than suffocates us due to the dominance it can achieve through hyper intelligence. We can only secure a human-centred digital future if we design ourselves into that world. My own personal philosophy is: 'Design your future or it will design you.'

My co-author Elizabeth provides the roadmap and the checklists that are as vital as the vision I have tried to express. Boldness of vision is crucial (or why bother making the journey), but without a map and clear signposts we become quickly lost, uncertain and confused. Leadership in realizing the opportunity of the Digital Renaissance of Work must remain visible and relevant, and when leaders get that right (as Richard Branson and Marc Benioff have), the experience of persistent change and development becomes thrilling rather than demoralizing. Work matters to each of us and always will in my estimation; it is a core aspect of a life well lived. So, now, in Part 2 of the book, let's see how to make the Digital Renaissance of Work come alive in your organization.

Note

1   Turkle, Sherry (2011) *Alone Together: Why we expect more from technology and less from each other*. New York: Basic Books.

PART II
Delivering Digital Workplaces Fit for the Future

Chapter 10

Your Digital Workplace Journey

As my co-author Paul Miller says, every organization already has a digital workplace. This is not a futuristic concept that will finally be reached when all the right technology services are in place. It is happening now – whether it enables employees to work productively and engages them, or is a source of problems and causes frustration – and it is evolving fast.

How would you describe the digital workplace right now in your organization as the Digital Renaissance gathers pace? For many organizations the digital workplace has evolved in a haphazard manner, creating a landscape that lacks integration and is difficult to use. The journey from this chaotic state to a cohesive, effective digital workplace needs to be taken one step at a time. This chapter is an opportunity to step back and assess where on its digital workplace journey your organization currently is. This assessment starts with getting to grips with the definition of the digital workplace and its promise for the world of work when aligned with business objectives.

One component of the digital workplace that can act as a rallying point is the intranet and the intranet team. Although the digital workplace comprises a much broader concept, the intranet can be a good starting point for understanding the wider set of technologies and how these need to be managed and developed. Already, advanced intranets offer a point of integration for many of the tools and services available in the digital workplace and, in some cases, the intranet team is evolving its responsibilities to take in a broader set of digital workplace tasks.

At the core of this chapter is a model for understanding your current digital workplace capability and its desired future state. The model is intended to be used by digital workplace professionals, such as IT strategists, workplace transformation leads, communications directors and intranet managers. It focuses on the totality of the employee experience rather than isolated elements such as intranet or social media tools. Its primary purpose is as a thinking tool for developing future strategy. In Chapter 14 we explore how to measure the progress and performance of your digital workplace in greater detail and on an ongoing basis, but at this stage the assessment is ideal for getting a quick primer on digital workplace capability.

In brief, this chapter looks at:

• What is the digital workplace?

• The promise of the digital workplace.

• Why are digital workplaces getting attention?

• The digital workplace dynamic in IT.

• Understanding digital workplace capability.

• Organizational readiness.

What is the Digital Workplace?

The simplest digital workplace definition is exactly what it says on the label: the digital environment of work that is not physical.

We can expand this to encompass the collection of all the digital tools provided by an organization that allow its employees to do their jobs.[1] This includes: intranets, unified communications, microblogging, HR systems, email, mobile applications, collaborative spaces, supply chain and customer relationship management (CRM) systems.

In the introduction to this chapter, I indicated that in many cases the intranet is acting as a rallying point for understanding this technology landscape. It is not that intranets are evolving into digital workplaces – a common misconception in our industry at present – but rather that intranets form a 'core component' of the digital workplace. This starting point for defining the digital workplace may be accurate but it in fact encompasses a great deal more than just a set of technology services.

When Paul started using the term digital workplace back in 2009, it had mostly been gathering dust in a drawer having been coined by Hewlett Packard in 1998 as the name for a printer.

In his book *The Digital Workplace – How technology is liberating work*,[2] Paul sets out very clearly why this term resonates for him:

> *We all know and understand physical workplaces as they have developed since the Industrial Revolution of the late 1700s – and the digital workplace*

is the 'digital counterpart' – the technology-enabled environments where work increasingly happens.

Viewed in this light, the digital workplace can't be defined in such simple and narrow terms as the component parts that can be described.

Historically, work has been defined by its location – in agricultural times, it took place in the village, fields and towns; in industrial times it moved to factories, cities and offices. Now, during the Digital Renaissance, work has become mobile, portable; not a physical location but a digital one. This means that the digital workplace is transforming work itself and how organizations create value.

A good way to better understand the digital workplace is through the impact it is having on how we work and live:

> ## DIGITAL WORKPLACE EXPERT VIEW
>
> *Digital workplaces exist. They're as real and relevant as physical offices. But most organizations aren't set up to manage them, instead focusing on individual technologies and some bilateral integration. Imagine the same principles applied to physical offices. What if there were no 'Facilities Department', just a 'Rugs Department' and a 'Doors Department' and a 'Chairs Department'? What if there were no architects or designers who worked together to create a complete physical environment? What if every door in an office looked different and required a different security fob to enter? Most modern offices are designed cohesively, but digital workplaces aren't. They aren't even seen as a unified environment.*
>
> Ephraim Freed, Communications Manager
> at DWG

- Work is no longer a destination – it can happen at the airport, at home, on the train, or in the office.

- Work is no longer constricted to 9 to 5 – work and life are blending, bringing both advantages and disadvantages to employees.

- Work is measured by output rather than presence – shifting performance management to a new dynamic.

- Work is becoming more satisfying – through the greater autonomy enabled by digital working, which is creating the new 'digital work ethic'.

- Work is disappearing for some roles, but new work and even industries are being created.

- Work is (in many cases) unfettered by geography – with remote workers and teams collaborating across time zones.

- Traditional concepts about trust at work are being challenged and then forged anew.

In reality, the digital workplace means different things to different people; you remember the Dove assistant marketing manager and the Verizon field engineer from Chapter 1? Their digital workplace needs are quite distinct. And this is how it should be. A salesperson on the road travelling from meeting to meeting has different needs and demands from an office worker who needs to stretch their day to work at home; or a customer service agent working at the frontline of the company; or a field worker on an oil rig. But the commonality is that the digital workplace is freeing them from outdated constraints and outmoded notions of what work is and where it happens.

The Promise of the Digital Workplace

As well as fundamentally changing the nature of work for us as individuals, the digital workplace is also challenging the traditional notion of how organizations organize, function and thrive. Paul has set out in vibrant colour, in Chapter 1, the historical and cultural context for this shift.

The promise of the digital workplace is becoming evident across both the public and private sectors, and in the most diverse industries, with digital workplace programmes varying in shape and size from agile working programmes, to Skype media walls, to augmented reality on the factory floor:

- Manufacturer ITAMCO digitizes its quality management system using iPads on the factory floor – and in the process reinvents its business model by becoming a supplier of specialist mobile apps within its industry.[3]

- The University of Liverpool transforms its intranet into an app store that enables staff and students more easily to get information, organize their schedules and stay connected from whatever device they choose.[4]

- Through a holistic Agile Workplace programme, Unilever achieves deep cultural change, removing artificial barriers around how work gets done and enabling employees to choose when and where they work, as long as the needs of the business are met.[5]

- An augmented reality app at Mitsubishi Electric provides repair operators with step-by-step instructions and indicators to assist machine repair.[6]

- Google glasses, currently being tested for engineers at General Electric, will transform work in the field by livestreaming knowledge to help diagnose and repair machines more effectively without the need to reach for a handheld device.[7]

- Federal agency, the United States Patent and Trademark Office (USPTO) transforms the lives of its employees by improving work flexibility for over 64 per cent of staff.[8]

- Alaska Airlines replaces 5,500 pages of paper captured in three ring-binders with iPads for their 1,500 airline pilots – not only lightening their load but giving them flight manuals, maps and the latest information via a range of critical apps.[9]

- Restaurant chain Tony Roma's uses QR codes throughout their kitchens to provide employees with on-the-spot training via their own smartphones.[10]

- Law firm Tredway Lunsdaine & Doyle LLP uses always-on Skype media walls to connect two remote offices for fully interactive meetings and social events; technology is enabling them to retain a family culture as the firm grows.[11]

- Speciality insurer Chubb uses a social business platform to democratize innovation within the company, leading to reinvention of its business model and processes.[12]

The list could go on and on. It is hard not to be excited by the sheer creativity and diversity of opportunities offered by digital workplace transformation. While industries such as high-tech, finance and retail are leading the digital charge, the promise of the digital workplace is becoming evident across all industries to varying degrees. And while some are transforming faster than others, no industry can afford to ignore the promise, as well as the impact, of digital working. From engaging the best talent (or losing it) through to reducing real estate (or paying the price), the potential impacts are very real. In Chapter 11 we will look at how this promise is translating into significant benefits for the organizations pioneering the digital workplace.

Why are Digital Workplace Programmes Getting Attention?

In DWG's research into digital workplace programmes, Julie Lakha, Head of DWG Consulting, highlights that it is no surprise that, with such rich rewards being reaped by digital workplace pioneers, the digital workplace is beginning to attract significant attention in organizations.[13] Senior management realize that, in order for their organizations to do more than merely cope with the ongoing wave of (potentially transformational) digital disruption, a new approach is needed. Digital workplace programmes offer such an approach. Pressure to adapt is now coming from both within and outside organizations, driven on by the ever-accelerating pace of digital progress.[14] The drivers include:

- cloud and mobile offering the promise of any device, any time, anywhere;

- higher expectations amongst all generations for digital enablement and connectedness;

- increasing difficulties around achieving work–life balance, information overload and a complex user experience for employees;

- the recognition that fragmentation of experience across digital tools is causing serious loss of productivity;

- the need to automate and streamline business processes;

- geographically dispersed teams – increased offshoring, outsourcing and mobile working have challenged connectivity.

- collaboration and social networking within, across and beyond the firewall;

- the recognition that the digital workplace is an essential enabler for wider agile working or workplace transformation programmes.

Organizations have concentrated on these areas in varying degrees, with investment and focus typically established across separate programmes to drive capability. While this has enabled the delivery of specific outcomes, it has also led to individual visions and strategies, which may or may not correlate. Consequently, there is growing acknowledgment that this fragmentation is detrimental and recognition that, in order to deliver increased business value and a cohesive employee experience, an integrated digital workplace programme with an overarching vision and strategy is needed.

The Digital Workplace Dynamic for IT

Disruption and reinvention are words that have become synonymous with the transformation IT is undergoing under pressure from the disruptive forces of social, mobile, the cloud and big data; while enabler, integrator and connector are words now synonymous with the ideal new IT department, replacing traditional notions of IT as a gatekeeper or service provider. The pressure to adapt is growing, as employees turn up for work with digital tools that may well be better and slicker than those IT are currently providing, and individual departments solve problems quickly using cloud tools.

In the old world, IT achieved simplicity by tight central management of technology services.[15] Despite this centralization, services tended to be managed within silos in the IT set-up, in order to keep things straightforward and to ensure that key services kept running. However, simplicity for IT has resulted in increasing levels of complexity for users as they attempt to work across multiple systems and devices. This can also be an issue where, for example, a local department has brought in software without understanding the impacts for users. Shortfalls in user experience are made up for by user effort, while cost savings for IT or an individual department are thrust onto the user in the form of frustrations and lost productivity. In reality, the breadth of these issues is not easily addressed. Moving towards a more user-centric view (understanding how people work, what motivates them and what their needs are) is a starting point though. Building on this, a more holistic view of systems can start to coalesce, challenging the old IT silos.

Digital units are springing up – either within IT or integrated across IT and business – and with them has come the rise of the Chief Digital Officer (CDO) to lead these new units. Poor relations between IT and the business are becoming outmoded, as Marketing or Communications and IT work closely together to deliver digital benefits, with this new approach seeking fusion rather than merely alignment.[16] The basic premise here is that the organization's digital estate (including its digital workplace) is not just an IT project or initiative. The vision is broader: a cohesive digital environment rather than focusing on individual services. This is about technology enabling change rather than technology change for its own sake. It is also about constant evolution: the digital workplace as a journey rather than a final destination.

DIGITAL WORKPLACE EXPERT VIEW

There are walls everywhere: 'This is the intranet, that isn't the intranet.' 'That's another team, we don't talk to that team.' 'This is our system, go away.' I think we can do this differently. We can't fix all of the problems of those muddled systems, but by bringing our attention up a level, changing our perspective and making some wise interventions, we can begin to bring clarity.

Chris Tubb, Benchmarker and Researcher at DWG

Understanding Digital Workplace Capability

Digital maturity has been shown to be critical to financial performance across all industries, both internally and externally.[17] Being able to assess this maturity – in our case, focusing on the internal digital workplace – is key to understanding the current state and desired future state, in order to help shape the digital workplace programme.

DWG's full digital workplace maturity model[18] takes a widescreen view of the digital workplace, assessing its user experience, strategic management, measurement and organizational context, as well as capability. In this section, we use a trimmed-down version of the model[19] to enable immediate insight into where your digital workplace is now in terms of its capability to meet core employee needs. It is a heuristic tool to help you gain understanding of the digital workplace capability in your organization. The model emphasizes the usefulness of the workplace as a whole, rather than specific features such as 'travel booking' or 'collaboration spaces'.

We break digital workplace capability down into five aspects:

- communication and information;

- community and collaboration;

- services and workflow;

- structure and coherence;

- mobility and flexibility.

LEVELS OF MATURITY

For each of these areas of capability we define five levels of evolution, from 'Base' to 'Excel', to define and codify the experience of employees and leaders across all five dimensions of the digital workplace (see Table 10.1).

The aim of the model is to be a thinking tool to help you understand where your digital workplace currently is and what options you have for advancing it. The model is not meant to imply that all organizations should seek to reach the 'Excel' or even 'High' levels in all areas. Rather, it allows you to consider what level of capabilities fit with your organization's strategy, employee needs, scale and culture. Furthermore, an organization's digital workplace often evolves along different lines. For example, some will emphasize online services or collaboration early on, whereas others will focus on tightly integrated communication and information provision.

Table 10.1   **The five levels of the digital workplace model, illustrated by typical employee and leader reactions**

			EMPLOYEES FEEL	LEADERS FEEL
1	Base	Entry level, typical of an intranet or collaboration tool when it first comes into existence.	'It's not relevant to me.'	'It's not relevant to me.'
2	Low	Some attempts at improvement, but offerings are still peripheral to the business.	'I use it when I have to, but it can be safely ignored.'	'It has no strategic relevance. It's something IT or Comms does.'
3	Mid	Relatively mature, but with room for improvement.	'It's mostly useful, but can be frustrating.'	'It's a practical tool, but I don't often get involved with it.'
4	High	The highest level an organization would normally expect to reach.	'I couldn't do my job without it.' 'Part of it is mine.'	'It's very important to how we operate, and I support it.'
5	Excel	A level of maturity beyond the norm. Strategically important to some organizations, but not necessarily to all.	'It's rewarding to use, and my needs are well anticipated.'	'It's made a significant difference to how we work.'

Digital Workplace Capability Areas

The following tables (10.2 to 10.6) outline the typical characteristics of an organization's digital workplace at each level. Note that these are examples rather than exhaustive requirements. Not all the characteristics may be true of your own organization, but they should enable you to find the closest match. When looking at the characteristics, consider how true they are for your total employee base (that is, not just workers who are currently office-based).

COMMUNICATION AND INFORMATION

What role does your digital workplace play in internal communications and as an information-sharing tool?

Table 10.2   The five levels for 'communication and information'

COMMUNICATION AND INFORMATION	
Base	**Static information storage** • The intranet is mostly used as an online store for static information such as policies. • News is updated sporadically and employees mostly get communication through other channels.
Low	**Top-down activity; static periphery** • The intranet is actively used for news, but it tends to come from the centre. • Employees do not see the intranet as their main communication channel. • There is a wide array of content, but much of it is outdated. • There is no assessment of the value or quality of content. • Departmental or local sites are mostly static (e.g. part of a document management system).
Mid	**Multiple, managed communication levels** • Most employees see the intranet as the place to go for regular news. • Some employees use online tools for two-way communication and feedback, but there are only a few examples. • There is some quality and value control around content . • Both global and local content are actively managed. • Online news is both centrally and locally produced. • Email is used primarily for local announcements only. • There is ad hoc coordination of communications teams.
High	**Structured, flexible content and communication** • Most employees prefer the intranet for nearly all communication and information needs. • Most employees feel that the intranet is a place where they can contribute news, opinion and information. • There is a broad mix of corporate, department, team and user-generated content with clear boundaries. • A wide range of media is used, including video and audio. • There is quality control appropriate to each level of content. • Content is not duplicated and there is clear ownership. • It is clear who publishes what to whom, and information is structured by audience not provider. • The user experience is personalized and customizable.
Excel	**Communication and content owned by all** • The majority of employees are both publishers and consumers. • User-generated content covers all media types (e.g. video, applications). • All employees understand the different options for using the digital workplace as a communication tool. • All employees are skilled in writing online content.

Basic capability here describes the traditional notion of a 'top-down' intranet that pushes out news and information via a central communication function – content that may or may not be up to date, relevant or of high quality. At the higher levels of capability, the intranet has become a powerful and dynamic tool for two-way communication and targeted, trusted content. High standards of content production are embedded in the organization's culture, and all employees are both consumers and publishers of a wide array of media types.

For many organizations across industries, evolving this capability has already been an important agenda item for communication and digital teams. As result, a base or low maturity in this capability should certainly be considered a major weakness.

COMMUNITY AND COLLABORATION

How well does your digital workplace support peer-to-peer working, including collaboration as a project team or community of practice, and social connectivity (such as finding people, seeking knowledge and sharing ideas)?

A low level of capability here is indicated by email being the only channel for online collaboration, and limited internal connectivity through a basic but often incomplete and unreliable people-finder. As the organization matures in this area of capability, a more complete employee directory will be put in place and instances of sporadic project collaboration and limited social network use may spring up in isolated pockets of the organization. If these indicators are noted and the impetus for new ways of collaborating and connecting across the organization harnessed, a more cohesive picture may start to emerge.

At the higher levels of capability we see people collaborating and connecting more seamlessly, using integrated and organization-wide tools that enable real-time access to colleague status, skills, projects and interests. Collaboration is deeply woven into how people work – both within and beyond the firewall.

For many organizations, this is the nut they are currently trying to crack, or at least they have recognized its importance and are formulating a plan of action. DWG research into collaboration within major organizations[20] has shown that, although the majority have now deployed collaboration and social tools of some kind, the results are often disappointing with cultural difficulties, lack of management support and weak governance all key factors underlying this.

Table 10.3   The five levels for 'community and collaboration'

COMMUNITY AND COLLABORATION	
Base	**No specific support** • No specific collaboration support – email is the main tool. • No, or partially complete, people-finder and locally maintained contact lists.
Low	**Ad hoc use of collaboration tools** • People-finder is mostly complete but unreliable, or there are multiple people systems. • Most collaboration is via email and shared drives, perhaps with some niche tools for team collaboration. • Tools in use may overlap in functionality or be 'unofficial' (e.g. Yammer accounts set up without IT's knowledge).
Mid	**Wide usage of disconnected tools** • There is a single address book with contact details, including some long-term contractors. • Collaboration tools are widely used for basic tasks such as document sharing and messaging but are not joined up. • Enterprise social network tools are in use by some groups but are not widespread or joined up (e.g. requiring a separate login for each).
High	**Online collaboration as a way of working** • There is a comprehensive directory of personal profiles, where people maintain their own information about skills, interests and social networks. Contractors and partners are included. • There are activity streams that can be followed for people and information (e.g. projects, documents or image libraries). • Private collaboration spaces are widely used (e.g. by project teams). • Communities are widely used for knowledge sharing and collaborating. • Integrated real-time collaboration tools are routinely used (e.g. presence, IM, desktop video and web conferencing). • Employees are supported in developing skills and techniques for using these tools. • There is a programme to cultivate employee adoption of these tools.
Excel	**Seamless collaboration outside and in** • There is permeability with the outside; employees routinely collaborate with third parties through the extranet and other secure environments. • Collaboration and social tools are fully integrated. • Immersive collaboration environments are commonly used, such as telepresence or virtual worlds.

SERVICES AND WORKFLOW

How effective is your digital workplace at delivering online applications that support employee self-service, workflow or specific functions, such as CRM or supply chain management?

The ability to deliver online services and workflow is core to an effective digital workplace. To start with, automated and manual processes may exist side by side as the digitization of processes is gradually extended in both capacity and reach. As more processes come online and become more widely available, employees may begin to be able to do the majority of their key tasks online – but the experience may not be an easy or pleasurable one, involving multiple logins and inconsistent interfaces.

Table 10.4   The five levels for 'services and workflow'

SERVICES AND WORKFLOW	
Base	**No online services** • There are no online services, although information may be provided about services that are delivered offline.
Low	**Basic applications online; manual back office** • Employees routinely use one or two standalone applications online. • Some services may involve online forms that are manually processed after submission.
Mid	**Key services online** • Key employee services (HR, finance, IT, facilities and travel) are used online by most employees. • Some services are limited to groups of employees (e.g. not in all countries). • Manual processes still exist. • Most 'work' tools (e.g. dashboards) that people use are dedicated applications disconnected from each other.
High	**Services and applications used online by all** • All employee services are used by all employees online. • Applications have a consistent interface and single sign-on. • There are joined-up processes and workflow (e.g. a new employee process in HR triggers IT processes for user accounts). • Online workflow is widely used, even for local activities (e.g. departments define workflows for common team tasks).
Excel	**Employees adapt applications to needs** • Employees use the digital workplace to combine and integrate data from multiple systems (e.g. for dashboards). • Mash-ups are used to help visualize and combine data from internal and external sources (e.g. map overlays, custom apps).

At the higher levels of maturity, the digital workplace has become a 'one stop shop' for all of these applications with branded experiences, single sign-on and joined-up processes enabling a seamless experience. There is the ability to integrate and combine data from different systems and tools to produce dynamic and highly useful dashboards or mash-ups.

As with the 'communication and information' capability, base or low levels of maturity in this area should now be relatively unusual for digital workplaces in major organizations. However, although key processes may have been digitized, with accompanying reductions in transaction costs, there are often still many opportunities for enhancing employee adoption and productivity via better integration, reach and usability.

STRUCTURE AND COHERENCE

Is your organization's digital workplace managed as a cohesive whole? This includes the extent to which platforms are shared; sites and applications integrated; branding consistent; governance in place; and standards adhered to.

A low level of capability in this area is indicated by a plethora of sites and applications with no overarching framework and resultant difficulties in navigating across a disjointed digital landscape. As maturity develops, the main platforms start to be aggregated, and more centralized governance and standards emerge across the digital estate – but exceptions and anomalies are still common. Reaching this mid level of capability is often driven by the need to create a 'one company' culture across a global organization.

Reaching the higher levels of maturity in this area of capability (which imply a fully integrated digital workplace for all) is still a far-off goal for most organizations. These levels include characteristics such as consistent information architecture and interface, scoped search, single profiles and appropriate governance throughout – the kind of things many intranet managers dream of at night!

MOBILITY AND FLEXIBILITY

How advanced is the digital workplace in providing access regardless of location and device?

This is about the journey of the organization towards becoming truly virtual. This journey, for most organizations, began with a notion of access that was limited to desktop computers in offices, for knowledge-workers only. It gradually expanded to additional access modes such as VPN/laptop and kiosks in frontline environments. Greater mobility within the office (thanks to WiFi) and extended options outside the

Table 10.5   The five levels for 'structure and coherence'

STRUCTURE AND COHERENCE	
Base	**No structure** • There is no formal management of the digital workplace framework. • There may be multiple small intranet sites. • The intranet is not connected to anything else.
Low	**Disconnected sites, internally structured** • Multiple active intranets exist (e.g. for departments, business lines etc). • The central intranet may link to other sites or tools such as wikis, but there is no deeper integration. • Some key sites may be well governed within themselves, but there is no consistency between sites. • Local sites may not be accessible to people outside that region. • Search does not index between sites. • Little or no remote access (e.g. only by staff with laptops and VPN).
Mid	**Aggregation of principal platforms; some standards** • Many sites may share a single platform but a significant minority still sit outside. • There are standards to align look and feel, even if different platforms are used, but there are exceptions and anomalies. • Search is federated across existing sites. • Look and feel is aligned between key sites and tools. • Office-based employees use the intranet and tools other than email at least weekly. • Access from mobile devices is possible, but there are no mobile-specific designs (e.g. apps or mobile style sheets).
High	**Integrated digital workplace** • There is a consistent user interface throughout. • There is a consistent information architecture and metadata. • There is a single profile and login for all services and social network tools. • Applications are integrated behind one gateway interface. • Search can be scoped to any level (faceted search). • Nearly all employees use the intranet and most use it several times a day. • There is clear governance regarding what to manage and what to leave open to user-generated content. • Mobile use of the digital workplace is specifically designed for and supported. • Kiosk or home access is available to all employees without office PC access. • Mobile (e.g. phone or tablet) and employee-owned devices are specifically supported.
Excel	**Digital workplace for all** • Components of the digital workplace are adapted to specific use-cases (e.g. sales support apps on tablets for frontline staff). • The intranet is absorbed into other elements of the digital workplace. • Nearly all roles will incorporate the digital workplace in some form. • All employees use the digital workplace daily. • All employees use the same application for a given task. • Innovation of digital workplace features is managed and encouraged.

Table 10.6   The five levels for 'mobility and flexibility'

MOBILITY AND FLEXIBILITY	
Base	**Access from desktop in office** • Digital workplace services are accessed using a network connection within the organization's office. • A large majority of users access the digital workplace using a desktop computer. • Flexibility in out-of-office working is granted only as essential, to a small minority (e.g. sales team, field technicians).
Low	**Limited mobility with VPN/laptop; mobile email and calendaring** • Access to the digital workplace is possible using a laptop PC via VPN. • There is access to email and calendar using a mobile device for specific roles. • Kiosks are available to all employees without office PC access. • Flexibility in out-of-office working is granted to out-of-hours support staff, as well as extending working days for a large minority. • It is possible to connect a company laptop PC in different offices within the organization, including internationally.
Mid	**Widespread mobility, smartphones** • The majority of users can access the digital workplace using a laptop when out of the office. • The majority of users can access calendaring and email using company mobile devices and there is nascent use of the intranet and services using a mobile device browser. • Non-company PCs can be used to access the digital workplace and employee-owned devices can be provisioned for access. • Flexibility in out-of-office working is granted for convenient exceptions or to support parental care for the majority of employees. • Offices routinely provide WiFi access to allow users freedom of movement.
High	**Apps and computerization** • The majority of employees access the intranet and other services using a mobile device browser with suitable interfaces provided for display and context. • Mobile apps have been developed for content and application services. • ePost services are available to scan physical post and make it available digitally. • Home extranet access is available to all employees without office PC access. • Flexible working is granted at line manager discretion for roles that are not location dependent.
Excel	**Unified, any place, any device: the virtual organization** • The digital workplace is available on any device, from any location. • The digital workplace is aware of location and context, and proposes relevant choices. • There are seamless transactions between different modes of access: desktop, mobile, projection. • Flexible working is expected, encouraged and is the choice of the individual.

office (thanks to mobile devices) has gradually been enabled, although often only for some employees with limited flexibility options.

These tentative steps into mobile and flexible work options are now transforming into major strides as employees begin to access more and more of the digital workplace via company- or employee-owned devices, with enterprise app stores and extended flexible work options springing up. For some organizations, flexible work options are becoming the norm, with agile physical workplaces and seamless digital workplaces underpinning new modes of working.

The availability of the digital workplace, whether this be in the office, at home, on the factory floor or in a retail outlet, is currently a key area of focus for major organizations that are waking up to the enormous potential mobile and flexible work capabilities can offer.

Organizational Readiness

<table>
<tr><td>

DIGITAL WORKPLACE BENCHMARKING SNAPSHOT

When DWG benchmarks the digital workplaces of major organizations, one of the areas of capability and maturity it assesses is organizational readiness, or to what degree leaders, managers and employees are enabled to use the digital workplace as a way of working. Organizations at the early stages of their digital workplace journey may only go as far as policies and e-learning resources to support usage of available services and applications. As the digital workplace starts to mature, senior managers support and promote the digital workplace as a way of working; management by objectives is promoted; and training for new employees is built into onboarding processes. In a fully mature digital workplace environment, employees can work seamlessly using all aspects of the digital workplace; supporting policies are simplified and rationalized; and managers have the ability to effectively manage their teams, regardless of physical location.[21]

</td><td>

Assessing your organization's digital workplace against this model will provide a clearer picture of current capability and future areas for development. This understanding needs to be put into context with the level of readiness for digital working within the organization, which can be evidenced by the degree to which digital tools are integrated into the organization's overall ways of working.

Does getting up to speed on digital tools and policies happen as part of the onboarding of a new employee? Is the intranet the trusted source for corporate information? Are the leadership team visibly using the tools and making their digital presence felt? How easy is it for managers to manage remote staff? Is it obvious how digital fits into the core values of the organization? Questions like these help to establish an understanding of the readiness

</td></tr>
</table>

of the organization for the digital workplace. The process is also about assessing the mindset within the organization as regards digital modes of working: whether there is a readiness to adapt modes of working based on new digital capacities, or whether digital is viewed as something to be merely fitted around current ways of working. The latter isn't wrong but it will impact the extent to which the digital workplace programme can make a real impact.

Investigation may show that some parts of the organization are more mature in their use of digital working tools and practices. These leaders can be of great advantage to the whole digital workplace programme, providing best practice examples and benefit cases. As the rest of the organization starts to follow, healthy competition around digital working practices can be developed between departments or teams.

Chapter 10 Key Takeaways

- Every organization has a digital workplace – whether it enables employees to work productively and engages them, or is a source of difficulties and causes frustration.

- The digital workplace is the collection of all the digital tools provided by an organization that allow its employees to do their jobs.

- The digital workplace is changing how we work as individuals and the way that organizations operate.

- Digital workplace initiatives of varying types are happening across industry sectors – from agile programmes for knowledge-workers in finance, through to augmented reality apps on the factory floor in manufacturing.

- Digital workplace programmes are starting to gain significant attention from senior management.

- The digital workplace represents a whole new dynamic and way of operating for IT departments.

- The digital workplace is not an advanced intranet. But in many organizations the intranet is acting as a rallying point for evolving thinking around the digital workplace.

- Understanding current digital workplace capability is critical to making a business case and establishing a digital workplace programme.

- Using the model set out here will help you to think about current capability across five areas: communication and information; community and collaboration; services and workflow; structure and coherence; and mobility and flexibility.

- The level of organizational readiness for digital working within the organization should be understood before defining the digital workplace programme.

Notes

1   Tubb, Chris (5 November 2013) So what is the digital workplace anyway? Digital Workplace Group News: http://www.ibforum.com/2013/11/05/so-what-is-the-digital-workplace-anyway [accessed 26.03.14].

2   Miller, Paul (2012) *The Digital Workplace: How technology is liberating work*. London: Digital Workplace Group: http://digitalworkplacebook.com [accessed 26.03.14].

3   Bynghall, Steve (2014) Success with Enterprise Mobile: How tools for frontline employees drive value. Digital Workplace Group: http://www.digitalworkplacegroup.com/resources/download-reports/success-with-enterprise-mobile [accessed 25.03.2014].

4   As 3 above.

5   Bynghall, Steve (2013) Digital Workplace Fundamentals: The integrated approach. Digital Workplace Group: http://www.digitalworkplacegroup.com/resources/download-reports/digital-workplace-fundamentals [accessed 26.03.14].

6   Metaio.com (18 March 2013) Mitsubishi Electric MeView augmented reality maintenance enterprise application. Metaio AR: http://www.youtube.com/watch?v=iz4ykMn3UR4 [accessed 26.03.14].

7   Boulton, Clint (7 November 2013) GE wants Google Glass for Christmas. *Wall Street Journal*: http://blogs.wsj.com/cio/2013/11/07/ge-wants-google-glass-for-christmas [accessed 26.03.14].

8   As 5 above.

9   Lakha, Julie (15 May 2013) Screenshots: Alaska Airlines and rollout of iPads. Digital Workplace Group: http://www.digitalworkplacegroup.com/2013/05/15/screenshots-alaska-airlines-and-rollout-of-ipads [accessed 26.03.14].

10   Chester, Eric (21 February 2013) How Tony Roma's encourages workers to use smart phones at work. TLNT: http://www.tlnt.com/2013/02/21/how-tony-romas-encourages-workers-to-use-smart-phones-at-work [accessed 26.03.14].

11   Maio, Pat (2013) A law firm uses Skype to bridge the distance. Orange County Register: http://www.ocregister.com/articles/firm-538878-offices-attorneys.html [accessed 26.03.14].

12   Fitzgerald, Michael (2013) Redesigning innovation at Chubb. *MIT Sloan Management Review*: http://sloanreview.mit.edu/article/redesigning-innovation-at-chubb [accessed 26.03.14].

13   Lakha, Julie (2013) Setting Up the Digital Workplace Programme. Digital Workplace Group: http://www.digitalworkplacegroup.com/resources/download-reports/setting-up-digital-workplace-programme [accessed 27.03.14].

14   Most industry experts anticipate that Moore's Law (the assertion that computing power doubles every 18 to 24 months) will continue to operate for up to a decade. As Brynjolfsson and McAfee put it in

The Second Machine Age (2014): 'The accumulated doubling of Moore's Law, and the ample doubling still to come, gives us a world where supercomputer power becomes available to toys in just a few years, where ever-cheaper sensors enable inexpensive solutions to previously intractable problems, and where science fiction keeps becoming reality.'

15   Tubb, Chris (21 November 2013) Why do most organizations ignore the digital workplace? Digital Workplace Group News: http://www.digitalworkplacegroup.com/2013/11/21/why-do-most-organisations-ignore-digital-workplace [accessed 26.03.14].

16   Bonnet, Didier (26 August 2013) Do you have the IT for the coming digital wave? *Harvard Business Review*: http://blogs.hbr.org/2013/08/do-you-have-the-it-for-the-com [accessed 26.03.14].

17   Capgemini Consulting/MIT Center for Digital Business Global Research (5 November 2012) The Digital Advantage: How digital leaders outperform their peers in every industry. Capgemini Consulting: http://www.capgemini.com/resources/the-digital-advantage-how-digital-leaders-outperform-their-peers-in-every-industry [accessed 26.03.14].

18   Digital Workplace Group (2014) Benchmarking Evaluations: Objective, expert feedback. http://www.digitalworkplacegroup.com/membership/overview/benchmarking-evaluations [accessed 25.03.14].

19   Adapted from: Marshall, Sam (2013) From Intranet to Digital Workplace: How to evolve your strategy. Digital Workplace Group: http://www.digitalworkplacegroup.com/resources/download-reports/from-intranet-to-digital-workplace [accessed 26.03.14].

20   Bynghall, Steve (2013) The Art of Collaboration: Optimizing online collaboration for success. Digital Workplace Group: http://www.digitalworkplacegroup.com/resources/download-reports/art-of-collaboration [accessed 25.03.2014].

21   As 18 above.

Chapter 11

Making the Business Case

In recent years a critical mass of evidence has emerged around the very real value that the digital workplace can contribute to the organization. More than this, it has been shown that an effective digital workplace is essential to the competitiveness of the organization. This is true across industries, with organizations that are recognized as digital leaders outstripping competitors in profitability, revenue and market value.[1] DWG's research[2] has revealed that a powerful business case for investment in the digital workplace is now possible with solid answers to senior management concerns about the cost of investment in technology infrastructure.

While we focus primarily on the 'hard' financial benefits that can flow from strategically aligned investment in the digital workplace, we also fold in the so-called 'soft' benefits and ways in which these contribute to the business case.

The benefits in summary are:

- **Cost optimization:** Opportunities to rationalize and reduce operational costs through digital workplace programmes can flow from reductions in real estate, reducing travel through virtual meetings, and consolidating systems.

- **People and productivity:** Increasing productivity levels among staff is a key driver for investment in the digital workplace – as are better retention and reduced absenteeism. Also important here is a range of so-called 'soft' benefits such as improved employee engagement and better employee experience.

- **Business continuity:** In recent years, disasters such as the tsunami in Japan in 2011 and Hurricane Sandy in the US in 2012 have demonstrated the criticality of employees being enabled to work remotely in order for organizations to avoid shutdown. Many examples have emerged of successful business continuity in disaster situations, which have spurred on the case for further investment in the digital workplace.

- **Corporate social responsibility (CSR):** The digital workplace is having a significant positive impact on the environmental footprint of organizations, particularly through reduced travel and real estate.

- **Increased revenue:** There are now diverse digital workplace interventions that can be made to better enable sales and frontline staff to increase revenues. These can be as straightforward as better information provision, or as intricate as gamification initiatives designed to spur motivation.

- **Accelerating innovation:** Where the organizational culture is supportive and the right digital tools in place, innovation is accelerating at an exciting rate. Beyond the traditional boundaries of R&D, the digital workplace is enabling organizations to extend participation in innovation to the whole employee population and even further to partners and customers.

This list is not meant to be exhaustive – for instance, other benefits may be found in improving risk management or enhancing information findability – but it does constitute the primary areas for consideration in making the business case; in other words, the ones that will get the Chief Financial Officer's attention.

But before we delve into the detail of these benefits and look at a range of examples from major organizations, let's first consider one 'ingredient' that can make or break the digital workplace business case: strategic business alignment.

Strategic Business Alignment

Altogether too many technology projects fail to materialize promised benefits because the focus has been on tools and features instead of how the technology supports existing challenges and organizational objectives, as well as finding new ways to create value. Establishing an overarching digital workplace strategy and clearly aligning it with business objectives from the outset are prerequisites to demonstrating the business case and realizing the benefits of investment in the digital workplace.

A pertinent example from within the digital workplace estate of what can happen when there is a lack of alignment is collaboration, which is often treated as a platform decision rather than a potential solution to real business issues. Research from Gartner[3] has shown that, while social collaboration is now deployed in 70 per cent of organizations, there is only a 10 per cent success rate. It attributes this to a worst practice approach of 'provide and pray'.

This finding tallies with DWG's experience that organizations will generally achieve greater success the more collaboration is formalized, structured and focused upon different work-related processes, functions and groups. Less value will be gained if collaboration tends to be informal or vague in scope.

But this is not exclusively an issue for collaboration initiatives. Across the digital workplace, a lack of strategy and ownership may lead to a fragmented experience that is misaligned with the needs of the business as a whole.

Achieving this alignment may represent a project in itself that sets out to understand how the various components of the digital workplace work together to deliver the strategic objectives of the organization. In Chapter 13, we look in more detail at how this can be achieved, including securing stakeholder involvement and senior management buy-in.

As we saw in Chapter 1, the digital workplace spans all aspects of work across and beyond the boundaries of the organization and it is rarely possible to invest in all the elements at once. If the maturity assessment showed that your organization lags in certain areas, it's crucial to assess how important those areas are to organizational objectives. This will identify priority areas of the digital workplace for investment as well as areas where trade-offs may be acceptable.

For example:

- If an organizational goal is to create a sense of 'one company', investment in improving 'structure and coherence' would be key to assist with consolidation, as would 'communication and information'.

- If organizational change, such as restructuring, is a priority, then investment in 'communication and information' and 'community and collaboration' is important to help facilitate two-way dialogue between leadership and employees.

- If reducing internal administration costs is a key business objective, then investment in 'services and workflow' to enable employees to self-serve for transactions and information is important.

With this clear tie-in between what the digital workplace programme is trying to achieve and the strategic business objectives, the potential benefits we cover in this chapter are more likely to be achieved. Moreover, demonstrating this tie-in from the outset is more likely to gain the attention of the C-suite for the digital workplace agenda.

Cost Optimization

Significant cost savings are being realized through successful digital workplace programmes in both the private and government sectors. Agile working programmes are helping to rationalize costly real estate portfolios through new ways of working. Another key driver of savings in this area is technology consolidation, for example, content management software, along with the associated resources needed for support and development. Additional savings can be identified via automating processes and monitoring operations in real-time. Finally, the reduction of travel is a clear area of benefit.

CONSOLIDATING SYSTEMS AND PROCESSES

A digital workplace programme provides the opportunity, with the right governance in place, to span areas of the organization that have previously acted as silos, both in technology and business processes. Opportunities to consolidate separate systems and to integrate business processes (resulting in savings due to reduced licence fees and fewer resources required to maintain separate systems) can be identified, along with openings for streamlining processes.

Duplicated effort may occur through maintaining content in multiple systems or through helpdesk time needed to support users. Also, when systems are simplified, processes streamlined and duplication removed, a knock-on benefit can be the increased adoption of consolidated tools by staff, leading to an improved return on the investment made.

An 'as is' assessment of the digital workplace can be useful in identifying areas where business processes are disjointed across multiple tools, systems overlap and duplicated content exists. As well as system consolidation, savings in this area may also come from pooling IT resources, sharing knowledge among help teams, consolidating procurement or managing demand through central governance. Such an initiative can also bring clarity around the level of complexity employees are required to navigate in order to get work done and the potential loss of productivity this may lead to.

Examples

- **PwC** saved approximately £1 million as a result of introducing a global collaboration and social networking platform, and closing down separate applications. See Appendix 1 for more information.

- A key benefit reported from a server consolidation initiative at **Westjet** as the amount of data burgeoned was a 30–50 per cent reduction in human resource costs in the data services team, equating to US$0.5 million annually.[4]

- **Shell** saw savings of more than US$1 million through the consolidation of country and regional intranet sites into one global intranet.[5] Savings were realized through reduced infrastructure needs and demand on IT resources. It has also had beneficial effects on information discovery through simplifying the content estate.

- The **Netherlands Defence Force** realized an 80 per cent reduction in annual running and operational costs by consolidating 50 intranets to one single portal.[6]

REAL ESTATE REDUCTIONS

Traditional offices are expensive, inefficient, inflexible and difficult to scale (particularly down) and evidence from various sources now shows that actual utilization of fixed desk space is generally poor. Through office hotelling, desk sharing, agile working programmes and the resulting changes to the office footprint, companies can dramatically reduce the capital drain of owning or leasing buildings. The US General Services Administration estimates that 36 externally mobile and interactive workers can be accommodated in the same amount of space as 24 deskbound workers.[7] In addition to this, there are savings on electricity, parking leases, travel subsidies, furniture, computer equipment, supplies, maintenance, security, insurance, taxes, communal area expenses, and environmental and safety compliance.

Investment in the digital workplace – and the accompanying change management to help employees work in new ways – is a prerequisite to enabling employees to work effectively while reducing office space. The cost savings from real estate reductions must be a key element in any business case for investment in the digital workplace. It makes it a much easier C-suite sell.

Examples

- The **United States Patent and Trademark Office (USPTO)** avoided US$19.8 million in new real estate expenses through agile working. It has leveraged homeworking and office hotelling to grow its workforce (from 6,000 to nearly 10,000) without increasing its real estate footprint. See Chapter 13 for more information.

- **Citi's** 'Alternative Workplace Strategies' programme to introduce agile working reported a 20 per cent cost reduction in real estate, with the return on investment (ROI) of implementation recouped within 8–18 months. Technology is identified as one of the critical factors behind the programme's success.[8]

- **GlaxoSmithKline** says it has saved nearly US$10 million annually in real estate costs by gradually shifting 1,200 employees at its Research Triangle Park, North Carolina office to unassigned seating. Similar moves outside the US have saved the UK-based company some £25 million (US$40 million) annually.[9]

- On any given day, more than 115,000 **IBM** employees around the world work in a non-IBM office. Forty per cent of the IBM workforce operates without a dedicated office space. The employee/desk ratio is currently 4:1, with plans to increase the ratio to 8:1 in field locations. IBM calculates that it saves US$450 million a year in reduced facility infrastructure and associated initiatives through agile working.[10]

REDUCING TRAVEL THROUGH VIRTUAL MEETINGS

The case for reducing travel through virtual meetings is well understood and, as options for remote meetings expand, new possibilities for cost savings emerge: from connecting remote teams or external partners through to executive conferences or company-wide 'jams' that happen via integrated collaboration and video tools. Research by Aberdeen Group found that organizations with 'best-in-class' video collaboration solutions reduced corporate travel by 22 per cent.[11]

As well as saving costs, the digital workplace has the ability to engage the whole organization in such events (for example, via live broadcasts and microblogging) rather than just a select few. An additional benefit for employees is reduced time spent commuting. And, of course, reducing travel through virtual meetings has a favourable impact on the organization's carbon footprint (see the section on CSR below for more on this).

Examples

- **NASA** saved US$21 million in fiscal year 2012 by replacing travel with video conferencing when possible.[12]

- **Cisco** reported a 75 per cent reduction in travel costs by bringing together more than 3,000 Cisco executives for a virtual rather than an in-person conference, using its own video and collaboration tools.[13] Cisco employees participating in a telework study reported cost savings of US$10.3 million a year in fuel that would normally have been used to commute.

- **Vodafone UK** cut travel costs by nearly one third by introducing voice, video, IM and collaboration software.[14]

- Global brewer **Heineken** reported savings of £250,000 in travel to meetings in its first year of using video collaboration.[15]

People and Productivity

An area of digital workplace benefits that has rightly received a lot of attention in studies to date is productivity. There is now a wealth of evidence to link productivity improvements with effective digital workplace investment.

Also falling within the sphere of people-related benefits are: improved employee loyalty; talent attraction and retention; and reduction in absenteeism. Here, we focus on productivity improvements and reductions in absenteeism and staff turnover.

> **DIGITAL WORKPLACE EXPERT VIEW**
>
> *According to a 2012 DWG survey, improved productivity of employees was regarded as by far the most important benefit to impress senior management. Some 70 per cent of respondents selected this as their first choice.*
>
> Nancy Goebel, Managing Director at DWG

Many of the 'hard' benefits we cover in this section arise out of having a happier and more engaged workforce thanks to the individual benefits to be derived from the digital workplace, including:

- more control over when and where work is done;

- better ability to blend work–life responsibilities;

- better utilization of whole skill set and ability to innovate;

- reduced costs and time expended on commuting;

- psychological and physical well-being.

These benefits depend on having an effective digital workplace in place, as well as a culture and policies that support flexible working. Without the latter, new ways of working may lead to employees who are working flexibly feeling stigmatized or isolated, which can be counterproductive. Developments such as the San Francisco Friendly Family Ordinance,[16] which mandates the right to request flexible work arrangements, and extensions to the UK Flexible Working Regulations[17] will hopefully encourage a move away from work cultures in which a certain stigma about new ways of working still exists.

PRODUCTIVITY IMPROVEMENTS

The facts show that the shift to a digital workplace has a clear positive impact on efficiency, productivity and quality of work – with marked financial benefits. Productivity savings should be prominent in any digital workplace business case.

The digital workplace impacts productivity in a number of ways. Agile work programmes supported by an effective digital workplace show a particularly strong impact on productivity, enabling work to happen wherever and whenever it needs to. Part of the rationale here is that the office may not always be the best place to get work done, partly due to continual disruptions. In a CDW survey of over 1,000 IT decision-makers in 2013, 74 per cent indicated that using tablet computers and smartphones had led to an increase in productivity in their organization, while 25 per cent of these said it had led to a significant increase.[18] In addition, time saved through not commuting is a win–win for both employees and the organization, with some of this extra time being devoted to work. There are also strong links between flexible working and engagement, with engaged employees being prepared to work harder and better.

In addition, the ability to deliver easy-to-use, integrated processes and systems has a significant impact on productivity. For example, when a new recruit joins the organization, there is just one workflow for the line manager to complete in order to set up their hardware, organize their security pass and grant access to the appropriate systems. We look further at the importance of creating these kinds of seamless experiences in Chapter 12.

Digital workplaces are progressively offering new opportunities for productivity enhancement within the organization. A range of wearable technology, including smart watches and glasses, will be at the heart of these enhancements. For instance, uses being explored for smart glasses include: point-of-repair diagnostics and advice for engineers; training and improved quality control for oil-rig workers; and instant access to radiology images during surgery for physicians. Gartner expects that smart glasses could save field service businesses as much as US$1 billion a year.[19]

Examples

- **Fife Council**, Scotland's third largest local authority, measured a 22 per cent increase in productivity in staff enabled to work flexibly using a mobile app platform that helps to streamline workflow. Overall, the programme is on course to save the council £20 million.[20]

- The **US Air Force's Central Adjudication Facility,** where 95 per cent of employees now work at home, saw a 55 per cent increase in productivity in one year of its agile working programme.[21]

- **BT**, a pioneer of agile working, now has 15,000 homeworkers out of 92,000 employees. The company finds homeworkers 20 per cent more productive and notes that they take fewer sick days.[22]

- **Alpine Access**, one of the largest all-virtual call centre employers in the US, attributes a 30 per cent increase in sales and 90 per cent reduction in customer complaints to its home-based agents.[23]

- Using a global collaboration platform for creating business proposals and tenders has led to a 50 per cent faster proposal time at **PwC**, with an 80 per cent reduction in issues around document version control, a common problem when email is used. See Appendix 1 for more information.

ABSENTEEISM DECLINES

People missing work through sickness and other issues is a major, but often unarticulated, cost to organizations. Absenteeism has proved hard to reduce to date. Many instances of staff being off work are due to factors others than actual illness that prevent them from working. The digital workplace produces significant and sustained reductions in absenteeism, and both the direct and related financial gains from this are dramatic. Research from the Chartered Institute of Personnel and Development (CIPD) shows that organizations are increasingly using flexible working as a way of reducing absenteeism.[24]

Employee absenteeism represents an enormous cost for businesses. Recent research by PwC[25] has estimated that absenteeism now costs UK businesses nearly £29 billion a year, while a Gallup survey estimated that US businesses experience more than US$153 billion in lost productivity each year through absenteeism caused by the poor health of their workforce.[26]

There are potentially many reasons why the number of sick days can be significantly reduced through agile working programmes. One of these may be that flexible working can result in working patterns that have a positive impact on employee health, thereby resulting in fewer sick days. For example, stress is a major cause of absenteeism.

Flexible work options can also help to tackle 'presenteeism', where employees come to work despite being unwell. A CareerBuilder survey of 3,700 workers found that nearly three-quarters (72 per cent) of workers typically go to work when they are sick. Workplace pressures and presenteeism may be causing workers to go in when they feel under the weather, as more than half (55 per cent) of workers said they felt guilty when they called in sick.[27]

Examples

- During a year-long agile working pilot in the **City of Ottawa**, employee absenteeism dropped by about 42 per cent.[28]

- A study by **Virgin Media** revealed that CIOs believe that implementing a flexible working policy would reduce absenteeism by 10 per cent, translating into a massive £1.7 billion cash injection for the UK economy.[29]

- UK insurance company **NHBC** has reported a low absenteeism rate (£328 average sick pay per employee, compared with £600 stated by the CIPD), as a result of its flexible working programme.[30]

STAFF TURNOVER DECREASES

Like absenteeism, staff turnover represents a significant cost. Some of this is a direct cost from processes relating to recruitment, onboarding and training, as well as due to potential lost productivity resulting from not having a full complement of staff. Here, even small percentage improvements can translate into big numbers financially year on year. In recent years, studies by leading organizations, such as the Society for Human Resource Management and the Association for Psychological Science among others, have found clear and unequivocal links showing that alternative ways of working have a positive impact on staff turnover.

There are also many softer issues, which are harder to measure, but nevertheless impact the organization. These include the potential for lost knowledge and expertise 'disappearing out of the door', as well as effects on the morale of the existing workforce. Not only do they see their colleagues leave but they may also have to take on an extra workload during the transition to a replacement.

There are a number of studies and surveys suggesting that alternative ways of working that give employees more flexibility and autonomy about how and where they work can lead to significant increases in employee satisfaction and loyalty. For example, Canada's 'Top 100 Employers' competition has found that employees who are given the option to telecommute report greater loyalty.[31] The CIPD has reported that 76 per cent of employers feel that implementing flexible working practices has a positive impact on staff retention.[32] There can also be a knock-on positive impact on productivity. Without the trek to the office (on average a 75-mile round trip for respondents in the Staples Advantage survey), 76 per cent of telecommuters are more willing to put in extra time on work and say they are more loyal to their company since telecommuting.[33]

Overall, alternative working patterns are incredibly popular among staff and may well prove to be a vital component of any strategy that seeks to keep employee turnover at a minimum, in turn leading to competitive advantage. This is an obvious element to include in any business case for investing in the digital workplace. And, of course, the impact of an effective digital workplace in retaining staff will also be felt in attracting new talent to the organization.

Examples

- Care service provider **RehabCare** estimates it is saving millions of dollars per year in job retention as a result of equipping staff with easy-to-use applications on iPhones and iPads for use at point-of-care, replacing laborious paper forms and questionnaires.[34, 35]

- For **BT**, the availability of flexible working arrangements resulted in improved retention, with the percentage (over the last five years) of its UK female employees returning to work after taking maternity leave reaching 96–99 per cent, saving around £5 million a year in recruitment and induction costs.[36]

- **Ryan LLC**, a tax advisory firm with 900+ employees, saw its voluntary turnover drop from 20 per cent to 6 per cent as a result of its MyRyan programme, which allows employees to choose where and when they are most efficient and effective. The percentage of those employees who plan on working at Ryan until they retire rose from 56 per cent to 85 per cent.[37]

Business Continuity

Employees are increasingly used to disruption to working patterns due to circumstances outside their control. While some disruption may be down to problems specifically at an organization, such as the company's headquarters being flooded, most of this 'disruption' is centred on the inability of an employee to be able to travel to a physical workplace, generally an office.

Many types of external factor can affect business travel and commuting. Occasionally these result from extraordinary situations (terrorist attacks, the cancellation of air travel due to volcanic ash in the air, steps taken to avoid a swine flu pandemic) but more often, they are simply down to poor weather conditions, for example, the cancellation of public transport or impassable roads.

Clearly, the ability of employees to access the various key systems and applications they need to carry out their normal work activities from any location is central to successful business continuity. In many respects, of all the benefits from investing in the digital workplace included in this report, 'operations continuity' may be one of the easiest to sell to senior management.

Generally, disaster recovery and business continuity processes are already well established within large corporates and government organizations, and are a clear agenda item for IT departments, although in practice the efficiency and depth of these plans may vary dramatically from one organization to another.

Of course, the bottom-line benefits of operations continuity only really come into play when a crisis actually happens and can only be expressed as the costs potentially saved, but as the case studies from the US Government (below) demonstrate, the costs savings can be substantial.

Certainly, if there is little or no provision for remote working during a crisis within your organization, then it is time to raise this urgently as an issue, and as a central argument for investing in a better digital workplace.

Examples

- **USPTO** was able to maintain productivity at over 70 per cent of its usual levels during 2012's Hurricane Sandy due to flexible working arrangements. Its call centre operation remained 100 per cent operational due to work-at-home employees. Approximately 64 per cent of USPTO staff regularly telework, compared to an average across federal government of 8 per cent.[38]

- Some 1,200 **US Defense Information Systems Agency** (DISA) employees worked at home during the February 2010 snowstorms, when most of the federal government was essentially 'closed'. And the **National Institute of Health** (NIH) finds homeworking especially useful for pandemic planning. During the February 2010 snowstorms, NIH kept working.[39]

- During the east coast snowstorms of winter 2010, the **US Marine Corp**'s Business Enterprise organization effectively maintained continuity of operations in the national capital region. Even though the federal government was officially closed for 4.5 days, Business Enterprise employees in the region were 'at work' for 63 per cent of that time. Homeworking accounted for 53 per cent of the work hours recorded.[40]

- The **US Government** revised its original estimate of the cost of lost productivity from snow-related federal office closures in 2010, from US$100 million a day to US$71 million, to account for the growing number of teleworking federal employees.[41]

Corporate Social Responsibility (CSR)

ENVIRONMENTAL GAINS

A standard component of any large corporate's external communications programme is now the CSR report. Inevitably, one of the key details featured in this will be the efforts the company is making to reduce its impact on the environment. Agile working programmes and investment in online collaboration tools that reduce the need for travel and real estate are often cited as some of the initiatives that contribute to this effort.

According to the Climate Group's Smart 2020 report,[42] currently the largest opportunity identified within 'dematerialization' (that is, reduced use of the world's resources) is homeworking. Although other dematerialization opportunities may come to prominence in the future, based on historic trends, the analysis found that homeworking would have the largest impact – up to 260 $MtCO_2e$ (metric tonne carbon dioxide equivalent) savings each year. For example, in the US, if up to 30 million people were able to work from home, emissions could be reduced by 75–100 $MtCO_2e$ in 2030, comparable to likely reductions from other measures such as fuel-efficient vehicles.

Reducing an organization's environmental impact makes sense on many different levels, with multiple benefits. Some of these are intangible, for example, the ability to help preserve or improve an organization's reputation, which can be particularly important if the main activity of the business has some specific environmental impact.

A more tangible benefit is the ability to reduce environmental taxes and penalties, which may have been introduced in different countries to help encourage energy efficiency or a reduction in carbon footprint. With carbon taxation and other environmental levies only likely to increase, active reduction in the carbon footprint of companies will become ever more commercially prudent, with a likely direct impact on the bottom line.

Examples

- **Microsoft** estimates that its employees avoided flying more than 100 million miles in the first year of using unified communication technology, saving 17,000 metric tons of CO_2.[43]

- **Capgemini**'s carbon emissions have fallen by 12.6 per cent since 2008 as a result of its TravelWell programme, which included providing technology alternatives to non-essential travel. It has also achieved WWF-UK's 'One in Five Challenge' (reducing business flights by 20 per cent in five years). A founder sponsor of the challenge, Capgemini achieved this in the first year, reducing flights by 4,508.[44]

- Through its award-winning flexible working programme, in 2009/10 **Ernst & Young** (now **EY**) avoided 6.8 million air miles by using video-conferencing facilities, while in 2010/11, it achieved a 24 per cent reduction in distance travelled by road per head and a 15 per cent reduction in CO_2 emissions per head, compared with 2006/07. It also achieved a 5 per cent reduction in distance travelled by rail and CO_2e emissions per head in 2010/11 compared with 2006/07. Its flexible working strategy has been supported by a £650,000 initial investment while the potential annual direct cost savings from business travel avoidance amount to £2.5 million.[45]

- In 2011, **E.ON** saved 25,668 hours of travel and 454.32 tonnes of CO_2 through the introduction of Telemeet (video conferencing over internet protocol; VCoIP) at 26 locations across the UK, achieving cost savings of £1,134,024. In the same year, it also saved 30,396 hours of travel and 991.13 tonnes of CO_2 through ongoing promotion and use of telepresence in these same locations, achieving additional cost savings of £1.8 million.[46]

INCREASED REVENUE

The ability of the digital workplace to promote an increase in revenue often manifests itself by enabling frontline staff, in particular the sales team, to better service and sell to customers. The availability of up-to-date information on clients and products, provided on devices that are mobile and intuitive to use, can help to maximize opportunities to cross-sell or upsell. We have already seen the impact that such initiatives can have on bottom-line benefits in the example from Barclays in Chapter 1.

As well as improving information and communication at the frontline, gamification techniques are helping to encourage healthy competition among sales teams to meet and exceed sales targets.

Examples

- **AT&T's** 'My CSP' knowledge portal includes a regular 'Retail Essentials Video' sent out to sales reps. When viewership rates were analysed against the attainment of sales goals, it was found that those who viewed the video three or more times a month had a 40 per cent higher sales goal attainment than those who did not.[47]

- The introduction to global employees of a collaboration tool, 'BUPA Live', based on the Jive platform, helped **BUPA** increase sales in the B2C Telesales & Telemarketing Group by 10 per cent, of which 'at least 1 to 2 per cent is thanks to the team supporting each other on BUPA Live'.[48]

- **HCL Technologies** introduced a sales portal, 'Wikiportal', based on the SharePoint platform, to connect over 10,000 sales, marketing and solutions employees spread across 31 countries in order to share ideas, knowledge and collaborate. One of its tools, the Reference Management System (RMS), maps prospects with existing customers to create a reference to help the closure of deals. HCL has found that the RMS increases the bid-to-win by 80 per cent compared to the company standard.[49]

- **LiveOps**, which runs virtual call centres, uses gamification (including badges, points and leaderboards) to improve the performance of its 20,000 operatives. Since introducing gamification, it has seen an 8–12 per cent increase in sales performance among certain agents.[50]

ACCELERATING INNOVATION

Just as the root of the word innovation means to renew or to restore, so the digital workplace is bringing renewal to the whole field of innovation within organizations. Innovation has traditionally been the stronghold of the R&D department. However, with the advent of digital tools bringing input and ideas from the whole employee population, as well as from partners and customers outside the boundaries of the organization, R&D is now far from the only source of innovation.

Technology can do little where there is a culture that discourages innovation. But where innovation is part of the organization's DNA, tools such as expertise-finders, ideas management systems and 'innovation jams' via the intranet are all accelerating and enhancing the process of innovation; and not just innovation around products and services, but also around business models, operational processes, productivity tools and more. According to a study by PwC, three-quarters of CEOs regard innovation as being at least as important as operational effectiveness.[51]

Where advanced employee directories or expertise-finders have been successful, they are enabling faster problem solving and sharing of expertise across the organization. There are also numerous instances of ideas management systems being effectively deployed on the intranet. These allow the whole employee population to contribute to organizational goals such as new product development, meeting environmental targets, or cost-cutting initiatives. Annual innovation jams, as championed by IBM, or more regular 'mini jams' can also bring a critical mass of attention around a particular issue or opportunity.

The opportunities for innovation enabled by digital technology can extend beyond the walls of the organization too, with platforms such as Affinnova and Kaggle enabling organizations to tap into the best talent and ideas through posing challenges and competitions to a global community of experts.

Examples

- **American Electric Power** introduced an ideas management system on the intranet, which is designed to enable employees to suggest solutions to real problems the company is facing. Within just a few months this helped the company to identify US$8 million savings.[52]

- Global building materials manufacturer **Cemex** was able to accelerate its product development cycle as a result of the introduction of its collaboration platform 'Shift'. Its functionality includes rich profiles that enable global connections and knowledge sharing around specific business problems, as well as an ideas management system.[53]

- An innovation jam at pharmaceutical company **Lilly** produced hundreds of ideas to help drive the new company values. Some of these have translated into practical solutions such as ways to meet environmental and safety goals ahead of deadline. One idea went on to save the company US$14 million a year through an innovative packaging approach. Other ideas have helped them to meet a range of corporate targets.[54]

Chapter 11 Key Takeaways

- There is now a critical mass of evidence that digital workplace initiatives can contribute very real value to the organization.

- It is essential to establish an overarching digital workplace strategy and clearly align this with business objectives in order to realize benefits.

- It is rarely possible to invest in all elements of the digital workplace at once; strategic alignment will help identify priority areas.

- Opportunities for cost optimization can be found through consolidating systems, reducing real estate and cutting back on travel; organizations such as Shell, Westjet, Citi, USPTO, NASA and Cisco have all realized major savings in these areas.

- Improvements in productivity, plus reductions in absenteeism and staff turnover, are being realized by organizations such as BT, PwC, Alpine Access and the US Airforce's Central Adjudication Facility.

- Disruption caused by various kinds of disaster is being minimized through digital workplace investment, with significant examples coming to light in recent years from US Federal Government organizations in particular.

- Leading organizations, such as Microsoft, E.ON, Capgemini and EY have reported significant reductions in their environmental impact through flexible working and virtual meeting capacities.

- Digital working is enabling increased revenue through better enablement of frontline staff to service and sell to customers; Barclays, AT&T and BUPA are examples of organizations leveraging the digital workplace in this way.

- A revolution is happening in the ways in which companies such as Lilly, American Electric Power and Cemex innovate; innovation now happens not just within the R&D department but across and beyond organizations, enabled by the digital workplace.

Notes

1   Capgemini Consulting/MIT Center for Digital Business (5 November 2012) The Digital Advantage: How digital leaders outperform their peers in every industry. Capgemini Consulting: http://www. capgemini.com/resources/the-digital-advantage-how-digital-leaders-outperform-their-peers-in-every-industry [accessed 27.03.14].

2   Miller, Paul (2012) Digital workplace business case. Digital Workplace Group: https://www. digitalworkplacegroup.com/resources/download-reports/digital-workplace-business-case [accessed 27.03.14].

3   Gartner (2 April 2013) Gartner says the vast majority of social collaboration initiatives fail due to lack of purpose. Gartner press release: http://www.gartner.com/newsroom/id/2402115 [accessed 27.03.14].

4   Lopez, Isaac (23 September 2013) WestJet flies lower after server consolidation. Enterprise Tech: http:// www.enterprisetech.com/2013/09/23/westjet-flies-lower-server-consolidation [accessed 27.03.14].

5   Intranet Dashboard (4 July 2013) iD and Shell profiled in The Wall St Journal. iD Blog: http://www. intranetdashboard.com/blog/2013/latest-news/id-and-shell-profiled-in-the-wall-st-journal [accessed 25.03.14].

6   Cisco (2007) Netherlands Defense Force merges more than 50 intranets in a single portal to communicate across all service branches. Cisco case study: http://www.cisco.com/web/about/ac79/ docs/wp/Netherlands_CS_0808.pdf [accessed 27.03.14].

7   GSA Public Buildings Service (2010) Leveraging Mobility, Managing Place: How changing work styles impact real estate and carbon footprint. GSA: http://www.gsa.gov/portal/mediaId/171183/fileName/ Leveraging_Mobility_508_compliant.action [accessed 27.03.14].

8   Citi (undated) Citi case study. Agile future forum: http://www.agilefutureforum.co.uk/wp-content/ uploads/2013/06/Case-Study-Citi.pdf [accessed 27.03.14].

9   Business North Carolina (2012) GSK plays musical chairs at work. Business North Carolina: http:// www.businessnc.com/articles/2012-06/regional-report-triangle-june-2012-category [accessed 27.03.14].

10 Global Workplace Analytics (October 2013). The bottom line on telework. Thurston Regional Planning Council: http://www.trpc.org/regionalplanning/thurstonheretothere/Documents/FINAL_ ThurstonTelework_formatted.pdf [accessed 27.03.14].

11 Aberdeen Group (November 2010) Telepresence and the Video Frontier. Aberdeen Group: http:// www.dsdinc.com/dsd/pdf/AberdeenWP-CommTelepresenceandVideoFrontier.pdf [accessed 27.03.14].

12 Ballenstedt, Brittany (2013) Agencies are saving millions with virtual events. Nextgov Wired Workplace: http://www.nextgov.com/cio-briefing/wired-workplace/2013/08/virtual-events-help-agencies-save-big/69161 [accessed 27.03.14].

13 Cisco (undated) Cisco Virtual Collaboration: How Cisco connected executives worldwide for strategic meeting. Cisco IT case study: http://www.cisco.com/web/about/ciscoitatwork/downloads/ ciscoitatwork/pdf/Cisco_IT_Case_Study_SLO_Global_Virtual_Event.pdf [accessed 27.03.14].

14 World Wildlife Fund (2011) Sustainability: Meeting the One in Five Challenge. Vodafone case study, www.org.uk/oneinfive: http://assets.wwf.org.uk/downloads/vodafone_challenge_case_study.pdf [accessed 27.03.14].

15 Polycom (10 July 2012) Polycom helps global brewer HEINEKEN save £250,000 in travel costs and 160 tons of CO_2. Polycom press release: http://www.polycom.com/company/news/press-releases/2012/20120710.html [accessed 27.03.14].

16 Sommer, Andrew J (28 October 2013) San Francisco gives employees right to request flexible working arrangements. Lexology – Green, Epstein Becker: http://www.lexology.com/library/detail. aspx?g=7396ad3f-fe11-4643-a387-1a6aafb446ee [accessed 27.03.14].

17 ACAS (2014) The right to request flexible working. ACAS: http://www.acas.org.uk/index. aspx?articleid=1616 [accessed 27.03.14].

18 Needle, David (16 May 2012) IT survey shows tablet deployment payoff is increased productivity. The Tablet Authority: http://tabtimes.com/news/ittech-stats-research/2012/05/16/it-survey-shows-tablet-deployment-payoff-increased [accessed 27.03.14].

19 Boulton, Clint (7 November 2013) GE wants Google Glass for Christmas. Wall Street Journal: http:// blogs.wsj.com/cio/2013/11/07/ge-wants-google-glass-for-christmas [accessed 27.03.14].

20   Rossi, Ben (4 November 2013) Fife Council on course for £20M savings from mobility deployment. *Information Age*: http://www.information-age.com/technology/mobile-and-networking/123457422/fife-council-on-course-for---20m-savings-from-mobility-deployment [accessed 27.03.14].
21   As 10 above.
22   BT (undated) Flexible working drives down more than costs. BT Insights and Ideas: http://www.globalservices.bt.com/uk/en/insights/more_productivity_from_your_force [accessed 27.03.14].
23   Lister, Kate (2010) Workshifting Benefits: The bottom line. Telework Research Network: http://www.workshifting.com/downloads/downloads/Workshifting%20Benefits-The%20Bottom%20Line.pdf [accessed 27.03.14].
24   CIPD (2013) 2013 Absence management survey. CIPD: http://www.cipd.co.uk/research/_absence-management [accessed 27.03.14].
25   PwC (22 July 2013) Sickness costs UK business £29bn a year and growing. PwC in Northern Ireland: http://pwc.blogs.com/northern-ireland/2013/07/sickness-costs-uk-business-29bn-a-year-and-growing-pwc.html [accessed 27.03.14].
26   Gallup (17 October 2011) Unhealthy U.S. workers' absenteeism costs $153 billion. Gallup Well-being: http://www.gallup.com/poll/150026/unhealthy-workers-absenteeism-costs-153-billion.aspx [accessed 27.03.14].
27   As 10 above.
28   CBC News (16 May 2011) Telework can save city money. CBC News Ottawa: http://www.cbc.ca/news/canada/ottawa/telework-can-save-city-money-councillor-1.1086549 [accessed 27.03.14].
29   Virgin Media (undated) Connectivity and employee collaboration – a match made in heaven? Virgin Media Business white paper: http://www.virginmediabusiness.co.uk/Documents/WP-CC-DD0713.pdf [accessed 27.03.14].
30   Business in the Community (2012) NHBC sustainable travel plan. BITC: http://www.bitc.org.uk/our-resources/case-studies/nhbc-nhbc-sustainable-travel-plan [accessed 27.03.14].
31   WORKshift (undated) Business benefits of remote work. WORKshift Benefits: http://www.workshiftcanada.com/workshift-business/benefits [accessed 27.03.14].
32   CIPD (May 2012) Flexible working and provision update. CIPD survey report: http://www.cipd.co.uk/hr-resources/survey-reports/flexible-working-provision-uptake.aspx [accessed 27.03.14].
33   Staples Advantage (2011) There's no place like a home office: Staples survey shows telecommuters are happier and healthier, with 25% less stress when working from home. Staples Advantage: http://investor.staples.com/phoenix.zhtml?c=96244&p=irol-newsArticle_print&ID=1586360&highlight [accessed 27.03.14].
34   Apple (undated) RehabCare: Post-critical care using iPhone and iPad. iPad in Business: http://www.apple.com/ipad/business/profiles/rehabcare [accessed 27.03.14].
35   MobileIron (undated) RehabCare advances mobile healthcare with iPhone and iPad. MobileIron case study: http://www.mobileiron.com/sites/default/files/case-studies/files/Rehabcare_Oct.pdf [accessed 27.03.14].
36   Department for Work and Pensions (2009) Flexible Working: Working for families, working for business. A report by the Family Friendly Working Hours Taskforce. http://www.dwp.gov.uk/docs/family-friendly-task-force-report.pdf [accessed 27.03.14].
37   Corporate Voices for Working Families (undated) Micro success story: Ryan, LLC. Ryan: https://www.ryan.com/Assets/Downloads/Articles/Micro_Success_Story.pdf [accessed 26.03.14].
38   Yoder, Eric (8 November 2012). Patent office says telework withstood Sandy. *The Washington Post*, DC. Politics: http://articles.washingtonpost.com/2012-11-08/local/35504910_1_patent-office-patent-examiners-uspto [accessed 27.03.14].
39   Overmyer, Scott P. (2011) Implementing Telework: Lessons learned from four federal agencies. IBM Center for The Business of Government: http://docs.caba.org/documents/IS/IS-2011-51.pdf [accessed 27.03.14].
40   As 7 above.
41   Marimow, Ann (23 March 2010) Cost to shut federal offices due to snow: $71 million. *The Washington Post*, DC. Wire: http://voices.washingtonpost.com/dc/2010/03/cost_to_shut_federal_offices_d.html [accessed 27.03.14].
42   The Climate Group (2008) Smart 2020 report. Global e-Sustainability Initiative (GeSI): http://www.smart2020.org/publications [accessed 27.03.14].

43   Department for Transport (November 2011) Alternatives to Travel: Next steps. Department for Transport: http://assets.dft.gov.uk/publications/alternatives-to-travel/next-steps.pdf [accessed 27.03.14].

44   Business in the Community (July 2011) TravelWell – Capgemini. ways2work award: http://ways2work. net/pool/resources/capgemini-2012.pdf [accessed 27.03.14].

45   Business in the Community (2012) Ernst & Young LLP – The journey towards sustainable travel and working. BITC case study: http://www.bitc.org.uk/our-resources/case-studies/ernst-young-llp-journey-towards-sustainable-travel-and-working [accessed 27.03.14].

46   Business in the Community (2011) E.ON – Sustainable travel at Sherwood Business Park and Electronic Solutions. BITC case study: http://www.bitc.org.uk/our-resources/case-studies/eon-sustainable-travel-sherwood-business-park-and-electronic-solutions [accessed 27.03.14].

47   Nielsen Norman Group (6 January 2013) Congratulations to the 2013 Intranet Design Annual Award winners! Nielsen Norman Group: http://www.nngroup.com/news/item/2013-intranet-design-awards [accessed 27.03.14].

48   Jive (undated) BUPA community hinges on having a single platform for communication and collaboration. Jive case study: http://www.jivesoftware.com/why-jive/resources/case-studies/bupa [accessed 27.03.14].

49   Microsoft (13 August 2013) HCL Technologies. Microsoft case studies: http://www.microsoft.com/casestudies/Microsoft-SharePoint-Server-2010/HCL/Wikiportal-Innovative-Sales-Collaboration-Portal-Impacts-Revenue-and-Productivity-Strengthens-Sales-Processes/710000002928 [accessed 27.03.14].

50   Enterprise Gamification Consultancy (updated 23 August 2013) Gamification facts & figures. Enterprise Gamification facts and figures: http://enterprise-gamification.com/index.php?option=com_content&view=article&id=41&Itemid=26&lang=en [accessed 27.03.14].

51   PwC (2013) Unleashing the power of innovation. PwC: http://www.pwc.com/en_GX/gx/consulting-services/innovation/assets/pwc-unleashing-the-power-of-innovation.pdf [accessed 27.03.14].

52   Robertson, James (10 December 2009) Gold winner: American Electric Power (USA). Step Two Designs: http://www.steptwo.com.au/columntwo/gold-winner-american-electric-power [accessed 27.03.14]

53   CEMEX (2010) The ROI of CEMEX Shift. CEMEX white paper: http://www.cemex.com/whatisshift/docs/CEMEXShiftWP_The_ROI_Of_CEMEX_Shift.pdf [accessed 25.03.14].

54   IBM (2008) Building a collaborative innovation culture. IBM Global Business Services: http://www.dtic.mil/ndia/2009USCG/WednesdayPanel5Washington.pdf [accessed 25.03.14].

Chapter 12

Designing for a Flexible Workforce

In Chapter 2, Paul Miller set out a vision of digital worlds that we want to work in; worlds that are 'beautiful, pleasurable and productive'. It is a compelling future and yet, for digital workplace professionals mired in internal battles to get even basic usability of systems acknowledged as important, this will seem a far-off dream.

Even aspects of the digital workplace that have been well crafted have usually been purposed for knowledge-workers on desktops rather than the increasingly mobile and flexible workforce that is emerging. Forrester characterized 29 per cent of the global workforce in 2013 as 'anytime, anywhere information-workers – those who use three or more devices, work from multiple locations and use many apps'.[1]

For most employees in most organizations, the experience of working across these multiple devices, locations and services is currently one of fragmentation, resulting in frustration and lost productivity, among other negative effects. This is the 'digital disappointment' that Paul speaks of in *The Digital Workplace: How technology is liberating work*,[2] which can repel the best talent away from an organization.

This poor digital experience speaks to the lack of priority that has generally been given to user-centred design in the enterprise space, both by vendors of enterprise systems and within internal development teams. The perception of user experience as a 'nice-to-have' reserved for public-facing websites is all too common, despite the fact that numerous studies have shown that dollars invested in user experience reap significant rewards in areas such as increased productivity and sales. For example, a 2011 study by the University of Texas showed that an increase of just 10 per cent in the usability of data for employees can translate into an increase of US$2.01 billion in total yearly revenue.[3]

On the other side of the equation, poor user experience is harming the bottom line of organizations through lost productivity, negative impacts on employee engagement, and the fact that expected benefits of digitizing processes are impossible to realize. Various studies have demonstrated the large amounts of time information-workers spend on online activities, such as searching for information, email and document-related activities.[4,5] When this is correlated with the cost per employee, the real damage to the organization of poor user experience starts to become apparent.

For example, one estimate, based on the average salary of a US Government worker, demonstrates an annual cost of US$8.4 million as a result of just 15 minutes per day of time wasted searching for information for half the employees in a 4,000 full-time equivalent (FTE) department.[6] In reality, time wasted searching for information in organizations may be as much as 20 per cent of total employee work time – and information findability is just one of the casualties of poor user experience.

To some extent, the issues of poor user experience have been hidden in silos within the organization as users struggle on with separate systems. However, the emerging digital workplace view is bringing these issues into focus as a more complete picture of the problems experienced by employees starts to emerge.

User experience considerations need to be at the heart of any digital workplace programme; in order to succeed it must be human-centred rather than technology-centric. Digital workplace professionals looking to address the issues of user experience are faced with a daunting task. Not only do they often have to contend with legacy applications that may have been introduced with little consideration of user experience, they also increasingly have to deal with off-the-shelf or externally hosted solutions offering limited control over the user experience.

At first sight, this may seem too big and engrained an issue to address. However, through a thorough understanding of the underlying issues, and by taking the necessary steps to address these through strategy, governance, standards and a toolkit of user experience interventions applied at the level of the digital workplace, a journey to a better user experience is achievable – albeit not overnight!

In this chapter we look first at gaining a thorough understanding of the issues inherent in designing a digital workplace that is fit for a flexible workforce:

- What is digital workplace user experience?

- What is the current experience of the digital workplace?

- The designed and non-designed digital workplace.

- The effects of poor user experience in the digital workplace.

- User experience debt.

We then go on to look at how to address these issues through:

- Digital workplace standards.

- User experience interventions in the designed and non-designed digital workplace.

- Working with procurement to prioritize user experience.

Understanding the Issues

WHAT IS DIGITAL WORKPLACE USER EXPERIENCE?

Let's remind ourselves first of the definition of the digital workplace from Chapter 10:

> *The aggregated set of technology services that enable us to do our work, including: intranets, unified communications, microblogging, HR systems, email, mobile applications, collaborative spaces, supply chain and CRMs.*

Even where the usability of particular tools or services within the digital workplace is looked at on an individual basis, the overall user experience is something much more all-encompassing.

There are many, varying definitions of user experience as it relates to digital environments.[7] Common to all of them are words such as 'ease-of-use', 'efficiency', 'fitness for purpose', 'simplicity', 'seamlessness' and 'attractiveness'; and these terms offer a good grounding for desirable user experience. They also speak to the subjective nature of the user's interaction with the product or service; the particular context in which this happens; and what it means to the user. These descriptions include more than just how easy it is to achieve the goal of an interaction, also taking into account how enjoyable and satisfying the process is, and whether the user is likely to want to have this experience again. Nielsen Norman describe 'a joy to use' as an ultimate goal for user experience.[8]

It is immediately clear that the digital workplaces of most organizations do not provide anything approaching a good user experience. Even where user experience has been thought about and designed at the level of individual services, it is unlikely the bigger picture of experience across the digital workplace will have been considered. From the perspective of their users, digital workplaces mostly consist of a jumbled mixture of commercial software clients and applications, operating systems, and a patchwork of browser-based tools, some designed with the users in mind and others off the shelf.

As well as describing the state of the user when using a service, user experience is also associated with a variety of design practices and techniques that have been devised by design professionals. These techniques are intended to result in services that are 'a joy to use'.[9]

However, these user experience practices have not been widely applied to designing the digital workplace as a single offering. Components of the digital workplace tend to be added piecemeal, project by project, and by a wide variety of business stakeholders. Some services – usually an intranet site or application – will follow good practices of designing with the needs of specific users in mind and testing with those users as the services are developed, but much of the digital workplace comprises commercial software developed elsewhere, meaning organizations are at the mercy of vendors' practices and competence when shipping products. Once such software has been installed there is often no way of changing it apart from waiting patiently for the next upgrade.

Digital workplaces are often derived through an almost feudal ownership from different areas of the organization and, as such, teams and systems can find themselves competing with each other for development resources and the attention of users. Changes to the user experience of systems are often resisted because of the problems this can cause when the system is upgraded, as is common with SharePoint.

THE CURRENT EXPERIENCE OF THE DIGITAL WORKPLACE

The visionary writer and computer scientist Jaron Lanier has pointed out the human tendency to try to make technology look better and more intelligent than it is: 'People degrade themselves in order to make machines seem smart all the time.'[10] In the case of the enterprise, employees may have little desire to make their technology look intelligent but they nevertheless have to create the impression of it functioning as a useful whole in order to get work done. In reality this usually means getting tasks done across a variety of disparate tools with varying branding, usability and so on.

The current experience of most digital workplaces is one of fragmentation. The level of disconnection between applications and devices that form the workplace landscape for most employees seems to be growing at an alarming rate. Investigations into this problem reveal statistics such as workers changing windows to check email or other programmes nearly 37 times an hour,[11] and that the top frustration for information-workers is that it takes more than one application to complete a simple task.[12]

Fragmentation appears in a variety of scenarios across the digital workplace, for example, disconnected collaboration tools and portals; multiple sources of information about employees across profiles and directories; lack of integration across document management systems and email; notifications across different systems; disparate branding across systems; and a lack of continuity and consistency across devices.

Fundamental Failings

The issues experienced by users in navigating the digital workplace are usually symptomatic of fundamental failings in how it has been developed and managed. These can include:

> ### DIGITAL WORKPLACE EXPERT VIEW
>
> *In user testing sessions our evaluators see users struggling with a proliferation of different windows as well as poorly labelled applications that offer no route back to the intranet.*
>
> *Common user tasks are often buried deep within the intranet structure and then users are expected to negotiate login barriers and clunky, unhelpful systems when they do arrive at the correct place.*
>
> *This results in a huge loss of productivity for organizations, where employees take far more time than necessary to complete common tasks like booking meeting rooms, requesting software and viewing holiday details.*
>
> Louise Kennedy, Usability Expert
> and Benchmarking Lead
> at DWG

- technology has been selected and implemented in isolation, either by a silo within IT, or within an individual area of the business such as HR;

- a lack of proper understanding of how employees work and different work modes across the organization;

- the absence of an enterprise information architecture;

- the perception at a senior level that user experience is a 'nice-to-have' rather than an essential part of internal tools and systems;

- a lack of balance between security requirements and ease of use;

- no clear and complete map of the organization's technology landscape;

- inadequate standards for digital workplace development;

- user-centred design techniques have not been applied to the digital workplace as a whole;

- the absence of overarching digital workplace governance.

The journey towards addressing these fundamental failings is one in which the strategic importance of the digital workplace and business case for investment (not just in technology but also the internal structures and processes required to focus it) must be acknowledged and articulated. We examined the business case in Chapter 11, and will go on to look at how to set up a digital workplace programme in Chapter 13.

THE DESIGNED AND NON-DESIGNED DIGITAL WORKPLACE

In DWG's groundbreaking research into digital workplace user experience,[13] from which many insights in this chapter are drawn, Chris Tubb highlights an important distinction between what he designates 'designed' and 'non-designed' areas of the digital workplace. This distinction is based on the level of control organizations can have over the user experience.

The 'designed digital workplace' is where organizations potentially have direct control over a system's interface and can therefore address a problem a user is experiencing when interacting with it. This is the world of the intranet and associated applications, mostly delivered via the browser, but can include any software tool the organization has the skills to adapt. All the tools and techniques in the field of user experience work in the designed digital workplace.

The converse is the 'non-designed digital workplace'. Mostly this is the operating system (OS), Microsoft Office, file servers, Microsoft Lync and enterprise software such as softphones. But the non-designed extends into the browser as well, with out-of-the-box SharePoint templates, Yammer and externally hosted solutions that might deliver services such as room and travel booking. In the non-designed digital workplace, organizations are at the mercy of vendors. Some commercial vendors of software take user experience seriously but many others do not. Either way, there is no way the organization can fix a problem with user experience directly and will need to look at ways to mitigate such problems with workarounds from within the designed digital workplace. For example, personalization on the intranet home page may be used to provide a relevant choice of intranet applications, which in themselves cannot be adapted. We will cover these interventions more fully later in the chapter.

Currently, there is a trend towards reducing the amount of designed digital workplace in preference for the non-designed digital workplace. Customization of software is viewed as being a risk, costly and an annoyance. Software-as-a-Service (SaaS) vendors, such as Salesforce, advertise 'No Software' as a benefit, offering a way to be up and running quickly. All digital workplaces rely heavily on commercial software,

DIGITAL WORKPLACE EXPERT VIEW

There are many reasons why users don't always voice their frustrations about their experience of the digital workplace, from cultural to a general feeling that nothing can be done. We frequently hear the employees who participate in our benchmarking evaluations telling us about all the workarounds and shortcuts they use to bypass difficult or unreliable systems.

Anticipating areas and factors that can cause poor user experience – such as maintaining a look and feel, avoiding complex and multiple logins, device compatibility, content and navigation restraints – can go a long way to presenting the digital workforce with a smooth and stable environment to operate within.

Louise Bloom, Usability Expert
and Benchmarking Lead
at DWG

particularly at the consumer application level, but pushed too far this can leave a large organization lacking the means to bring structure and coherence to a large tangle of functionality.

The effects of poor user experience in the digital workplace

To understand why good user experience should be the target state, we should look at some of the effects of poor user experience on both employees and organizations.

Poor user experience reduces the adoption of new services

The ultimate goal of any internal investment in technology is its adoption. Designing and implementing such services with the user in mind, and also using the techniques of user experience, will critically reduce the risk of these services not being adopted and, indeed, may speed up rates of adoption.

Poorly designed and implemented services provoke significant uncertainty in potential users, who may not understand the offered solution or may find it too complex, or, crucially, it might not be any more useful than the existing solutions. The result will be low adoption rates and a lack of return on the investment.

Poor user experience causes a loss of productivity

Looking at the whole digital workplace reveals a huge variety of tasks that can be accomplished, in innumerable combinations. In considering this, a broad view of productivity is helpful: the ideal digital workplace would support employees to do what they need to do, without confusion, interruption or meeting dead ends.

Productivity can be lost across a range of aspects of the digital workplace: having appropriate devices and tools to do the job; connectivity; support and training on tools; the ability to access information and people; and, increasingly, the ability to filter information and connections to avoid overload. Examples might be: having to boot up a laptop to find the details of your next meeting rather simply looking on a smartphone; the virtual private network (VPN) connection dropping; not being able to find WiFi when on the move.

Poor user experience annoys users and erodes employee happiness and satisfaction

Poor user experience can undermine employee satisfaction and happiness, both increasingly important indicators of employee experience and therefore productivity. Immediate frustrations with systems, such as time to boot, the need to key in repetitive information, and dropping connectivity or latency, can lead to dramatic swings towards negative emotions for most people. These feelings, of course, detract from an employee's task in hand, as well as reducing their good will towards the systems they have access to. Over time, mounting frustrations can lead to a drop in employee satisfaction, although specific causes may be difficult to surface in an employee satisfaction survey as the scope of questions tends to be very broad. For example, whatever the frustrations of using a digital workplace solution, both teleworking and autonomy over one's location have been found to increase satisfaction.

Poor user experience reduces the organization's ability to attract and retain talent

New hires coming into an organization where the technical infrastructure is poor and frustrating will question their choice of employer. Pride in the organization may also be an important factor in retention.[14] Regular research by Cisco suggests that the importance of technology to 'Gen Y' is not to be underestimated.[15] However, this is not necessarily restricted to digitally native 'millennials' and can apply to any new employee who has raised expectations due to consumer technology or who has been used to better enterprise technology in a previous role in another organization.

Poor user experience encourages risky behaviours

If users cannot achieve their desired tasks using the tools available to them, they will consider alternatives. If the user experience of parts of the digital workplace frustrates them, some of their chosen alternatives may well be, knowingly or not, in contravention of security and regulatory rules. Training, communication and stringent security will only go some way towards dissuading employees from adopting what they need and want to. The use of Dropbox within most organizations is forbidden, usually because Dropbox does not fulfil the requirements for data security of large organizations. That said, anecdotally, the use of Dropbox is exploding within some organizations as the simplest way to get files onto people's iPads. The provision of a cloud storage application that meets security requirements, and is as easy to use as Dropbox, would reduce this risky practice to a negligible point.

USER EXPERIENCE DEBT

Underpinning all of the above effects of poor user experience is the idea of 'user experience debt', a term used among user experience practitioners and put into the digital workplace context by Chris Tubb in recent research. The concept is derived from 'technical debt',[16] the notion that any shortfall in software development must eventually be repaid in the next round of development. User experience debt is the concept that the accumulation of poor user experience, or inattention to user experience across a system (or in our case an entire digital workplace), must be repaid somehow. Gaps in user experience left by digital workplace projects will be paid for by the time and frustration of the users.

Digital workplaces are the accumulation of many separate IT and business initiatives over decades. Each project will have been conceived in order to achieve certain requirements for a sponsor, while also attempting to hit targets of time, cost and quality, but this typically leaves gaps in the overall experience. If we attempt to imagine the collective experience of the gaps between systems, the complexity deepens even further.

User experience debt comes in two flavours. One form is incurred by not initially understanding the user's problem. In this emergent field we can sense and probe our way around this: we might not get it right first time, but by observing users we can probably improve their experience. The other kind of user experience debt is more pernicious: not caring about the experience and making it somebody else's problem. This might be to hit a deadline, come in at a given budget or because a bit of software came with an enterprise software licence agreement.

Costs may be hidden throughout the organization. For example, a department might save itself time by delegating some of the work it has previously done to colleagues in other departments. The originating department will save resources, but the total time spent on the task by employees in the organization as a whole may be far greater than that saved. Or reduced productivity due to a shortfall in the user experience design of a particular system may go unidentified, with additional resourcing brought in to try to mitigate the issues.

Addressing the Issues

DIGITAL WORKPLACE STANDARDS

Addressing the issues around the current experience of the digital workplace needs to be underpinned with a set of standards for the digital workplace. Elements of these may already exist at an individual system level, and these can be used as a starting point.

In many organizations, this starting point may be found in the governance and standards in place for the intranet. In DWG's research into the strategy and governance of intranets,[17] based on our benchmarking of intranets in major organizations, we found that in the period 2007 to 2013 there was a clear improvement in the standards and controls embedded within intranet governance structures. These include: standards for design, usability and accessibility; guidelines for writing; a content framework; and clear ownership of content.

However, this same research revealed that intranet teams still struggle with implementing effective controls for applications that are integrated to some degree with the intranet landscape, including achieving single sign-on. The difficulties experienced by intranet teams in these areas are indicative of the issues also faced in addressing user experience at the level of the digital workplace. DWG benchmarking of major organizations indicates an ongoing lack of: clear standards for the design, usability and accessibility of new applications related to the intranet; consistent user experience across new services and applications or a common authentication method; and processes to ensure compliance with the intranet strategy and standards.

DIGITAL WORKPLACE BENCHMARKING SNAPSHOT

In benchmarking the usability of digital workplace tools, DWG conducts an expert review and usability testing of a number of agreed tasks. These tasks are evaluated against criteria that look at content structure and integration, as well as: design; discoverability of content; ease of navigation; the experience of using the tools; accessibility; the effectiveness of tools for the relevant tasks; how easy it is to complete tasks first time; and the attitudes of users. For organizations scoring highly in these areas, their design and development processes are underpinned by clear standards that are properly enforced.

To an extent, intranet teams have been attempting to tackle the user experience of the wider digital workplace. As digital workplace teams and governance structures are established – either as an evolution of the intranet team, or around it – they will become better positioned to address these broader issues.

As the digital environments in which we work mature, we will increasingly expect them to be governed by a set of standards that inspire confidence through clear accessibility; appropriate branding; universal navigation and search capacity; quality content; user-centred information architecture; security and compliance. Just as we have come to expect certain standards in the way our built environment is constructed, we are coming to expect better user experiences inherent within our digital environment.

Good practices evidenced in the area of standards include:

- the adoption of a particular set of processes for application development that ensure alignment across different aspects, such as brand, legal compliance and IT strategy;

- the existence of strong guidelines for new web development, particularly relating to accessibility and branding;

- clear and up-to-date information about ownership on individual pages or applications;

- guidelines that set out standards for layout; use of colour palette and logos; writing for the web and other formats; plus appropriate review cycles to ensure compliance;

- using policies and standards that relate to physical workspaces (for example, retail environments or meeting rooms) as a starting point to develop policies for online tools.

Establishing standards for the digital workplace is an important step. However, while standards can help to improve quality, consistency and the security of the digital workplace, they cannot help define what is useful or what will be adopted by users. For this we need to look at applying user experience practices at the level of the digital workplace as a whole.

SOLUTIONS TO PROBLEMS OF USER EXPERIENCE IN THE DIGITAL WORKPLACE

Practitioners of user experience, mostly in the arenas of web, intranet and app development, have built up a variety of methods, tools and heuristics that assist in designing end products in ways that will hopefully be well received by users. As has already been described though, the experience of parts of the digital workplace may not be within the direct control of the organization. However, even then, there are still ways of mitigating problems of use and experience. This section will describe various methods and look at how these can be applied across the digital workplace.

The dream is to provide an integrated digital workplace that is perfectly adapted and designed to the needs of its users. This, in its strictest sense, is largely impossible as the digital workplace is too big and too costly to change in one go and much of it relies on off-the-shelf enterprise software. Some applications in the non-designed digital workplace are de facto standards; for example, it is unusual to find an organization that does not rely on Microsoft Office, Outlook and Exchange. Top-down design in the designed digital workplace is usually limited in scope and centred around the intranet, but sometimes excludes collaboration environments and often excludes intranet applications.

USER EXPERIENCE DESIGN AND INTERVENTIONS IN THE DESIGNED DIGITAL WORKPLACE

Good experience starts with a focus on the user and ensuring that, whatever the demands of internal stakeholders and the technology available, the solution remains useful and usable to the user. Some of the tools that can be deployed in the digital workplace are:

User experience strategy

A user experience strategy is a document that articulates how the user experience of a product fits in with the business and digital workplace strategy at the beginning of a project. It is helpful to document the current state of the digital workplace; to understand how the product will interact with other parts of the digital workplace; and to define some objectives for what the product will achieve.

User research

User research can take many forms. Development teams can look at certain demographic data, such as age, job role and geographical distribution, which may be sourced from HR records. Analysis of existing usage data from the intranet metrics platform may reveal insights. Questionnaires can be deployed to understand users' direct needs. Interviews with potential end users can take place in order to understand: the tasks they find most important; their frustrations with current solutions; their attitude to technology; and their demands for functionality.

Mental models

A mental model is a diagram that shows people's potential motivations relating to something, grouping people with similar motivations together. Evidence from user research into a topic can be collated to communicate users' motivations around it. Mental models can be used by development teams to make sense of different points of view amongst a user population and to help begin to segment potential users by their motivations and attitudes.

Personas

Personas arrange important user characteristics and attitudes, derived from user research and problems the project is trying to address, in a short pen-portrait of a fictional person. The advantage of using personas is that it allows project team members to empathize with the circumstances of the personas and therefore to dispel some of the preconceptions that may have built up through focusing purely on: technology ('This system does X'); project realities ('We only have seven developer days left and we have to ship it by next month'); or the demands of stakeholders ('The boss said he wants it like that'). When a question is posed, or a decision must be made, the project team can use the personas to understand whether this is likely to have a positive or negative effect on the end users of the technology.

User scenarios

User scenarios make use of developed personas and look at specific tasks and circumstances, plus the reactions of the fictional personas to those tasks. A user scenario map can then record the potential interplay between the user's task and the requirements for the end product.

User and stakeholder workshops

Participatory design is an attempt to closely involve stakeholders and users of a service in its design, as well as create advocacy and buy-in. Tools such as personas and scenarios, while they attempt to get inside the heads of the end users, are no substitute for asking and testing. Design sessions with end users make use of workshops in which thinking to date is shared, and users and other stakeholders, such as administrators and internal clients, share their vision for how it should work.

UNDERSTANDING EXPERIENCE THROUGH PERSONAS

The development of digital workplace personas gives us the opportunity to understand how people are working across the organization, as well as the frustrations and opportunities the digital workplace presents for them. Going through this process can also help to update outmoded views of desktop-centric work practices to a more holistic view of an increasingly flexible workforce.

At American Express,[18] employee work styles have been clearly categorized into four types: Hub, Club, Roam and Home. This categorization helps to define employee needs in terms of both the physical and the digital workplace. For example, 'Hub' employees require fixed desks, while 'Club' and 'Home' employees only need an office-based desk some of the time. There is also flexibility for these categories to be tailored somewhat to individual employee needs. In Appendix 2 we see how Cisco settled on five core work styles, which were then mapped across key business lines as part of the 'Cisco Connected Workplace'.

These are good examples of how user experience interventions – in particular user research and personas – can help to orient digital workplace development towards the real needs of employees. In DWG's 'Digital Workplace User Experience' report[19] Chris Tubb explored this approach with a set of detailed, illustrative personas based on extensive discussions with digital workplace managers and employees of typical organizations within large multinational corporations. These personas were intended to initiate discussions within digital workplace teams about the fit of their own major segments of employees. For these to be truly valid as user research, primary research needs to be carried out with real users at a single organization.

In brief the personas are:

Mobile in office: always on-the-go within the office and needs the digital workplace to be flexible and quick to access within the enterprise's own space.

The road warrior: always travelling on business and staying in hotels and needs to be able to connect quickly to the digital workplace any place, any time.

The field operative: mobile but within a defined region and checks into a company location at the beginning and end of every day; needs the digital workplace to organize work, connect with colleagues and manage their needs as an employee.

The day stretcher: mostly based at an office during normal working hours, but needs to be able to extend work into evenings and weekends at times.

The one-day-a-weeker: mostly based at an office during the week, but will work from home once or twice a week.

The homeworker: works mostly from home but will travel into the office once or twice a week.

The free agent: entirely independent in where work can happen and free to choose how and where work is carried out.

As well as investigating the way people work, the research also looks at the frustrations and opportunities within the digital workplace for each work style. For example, an Executive VP of HR with a 'mobile in office' profile would benefit from a low-latency device, such as an iPad, for quick access to budget approvals and communications in a hectic meeting schedule. Or a Quality Assurance Analyst with a 'homeworker' profile might benefit from online communities and forums to help avoid a sense of isolation when working from home. Or the Senior Engineer with a 'field operative' profile would benefit from home access to key systems rather than trying to access a PC at a depot that is shared with 15 other field engineers.

As well as being a good example of how user experience interventions can be applied at the level of the whole digital workplace, these sample personas also demonstrate just how varied the needs of employees of the digital workplace are.

Wireframes

A design must be documented and communicated to various stakeholders as it is developed and ultimately approved. Traditional project documentation is very text-heavy and potentially impenetrable to the lay-person. By documenting interfaces pictorially, using wireframe diagrams, it is easy to communicate what a system will do. Other deliverables can also be drawn: site maps show spatial organization; use cases show who can do what in a system; storyboards help to get a feel for the flow of the user's experience; and workflow diagrams show how documents or other assets move through various states in a workflow system.

Testing with users

Whatever people say a system should do and how it should be organized, there is no substitute for asking a representative user to use the system, then observing his or her reactions in order to ascertain whether they can indeed complete the task effectively. This can be done at many stages in the project's lifecycle: paper prototypes can be created from the wireframes and users asked to say where they would click; simple HTML prototype sites can be created (or exported from tools such as Axure) and put in front of users; a minimum viable prototype can also be tested and the design adapted. All of this is intended to avoid and minimize major changes late in the development process.

Interestingly, Xerox has experimented in this field by creating test labs based on non-traditional work environments. For example, so-called 'experiential prototypes' provided a test subject with the environment of a car, complete with a video screen to represent the view from the windscreen and a steering wheel.[20]

Iteration and intervention

The key to perfecting anything is to iterate its design when a problem is found. Problems may be reported by users or noticed by other means, such as direct observation or usage data. At this point, using any of the techniques described above can be employed at a small-scale maintenance level – user experience can be improved by using design as an intervention.

User experience interventions in the non-designed digital workplace

As already described, organizations do not truly control the experience of much of the digital workplace. Microsoft products, such as Windows, Outlook, Office and Lync, are still very prevalent on the desktop and SaaS products, such as Office 365, Yammer and Salesforce, are increasingly important within the browser. Enterprise mobility leverages VPN solutions from Cisco and virtualization technologies such as Citrix Receiver. There are some elements the organization can control, for instance, ensuring that capacity is managed correctly to optimize response time, and the configuration of brand assets such as templates, fonts and standard email footers. However, when problems arise and users can't find or configure something – and can't work it out for themselves – the numbers of calls to the help desk increase, as long as using the technology is essential; if it is just a 'nice-to-have' the users simply give up trying and don't bother to use it.

There are, however, ways in which such problems can be managed and mitigated that lie outside of the traditional tools of user experience practitioners. Otherwise, digital workplace managers just have to sit there and hope the user experience will be better in the next release. When something doesn't work correctly in the non-designed digital workplace, remedial actions can be performed in areas over which the organization does have control, that is, the designed digital workplace.

Structure: navigation, links and cataloguing

The intranet allows arbitrary linking to any resource. This can be used to superimpose structure on resources that have none; for example, by pulling together numerous intranet applications or providing a secondary option for navigation that is not provided by the other system. By creating a small database to manage records by their metadata, users can find useful resources, whether they know of their existence or not.

Personalization and customization

Personalization can be used to target the most relevant resources to the user, based on their profile; for example, showing only the intranet applications that user has access to and hiding the many hundreds they do not. Personalization in the digital workplace could also be used to prioritize content and resources based on current context or location; for example, home-workers would see one view while those travelling would see another. Customization can allow the user to prioritize their most useful resources.

Data feeds and APIs

Many systems that cannot have their experience redesigned have the ability to provide data feeds for use by other systems or offer an application programming interface (API) so that other systems can programmatically control them or interrogate them and display certain data. This provides options in the designed digital workplace to alleviate problems with findability or poor interfaces.

Enterprise search

Enterprise search is a great leveller of content because, as long as a resource is available via HTTP (hypertext transfer protocol), it can abstract and summarize the content, irrespective of formatting and source system, and it is able to do this autonomously and automatically.

Content

Published content, such as online training, manuals and knowledge bases for complex systems, is the first stage in assisting users to solve problems themselves. IT help desks usually monitor calls closely for trends that may indicate a problem.

Branding and visual design

The visual design and branding of off-the-shelf software is often viewed as a 'nice-to-have' but it also provides important cues to users that they are on a corporately supplied system. Good branding lends credibility to a system. If there is some degree of capability of branding within a browser application, navigational elements can be inserted to give the system a place within the hierarchy of the intranet that will offer some support to the user.

Working with procurement to prioritize user experience

Decisions about software tools are often made for strategic or cost reasons by people who are not the end users but senior executives in technical functions. This has a knock-on effect that the enterprise software products themselves are developed and marketed in order to appeal to the interests of senior decision-makers. In order to improve the base experience of enterprise software, organizations need to get user experience thinking into the procurement process. This could happen in the following ways:

- **User-centred requirements specifications.** Unless the thinking behind the specification for a system remains user-centred, and follows the kind of exploration described in 'User experience in the designed digital workplace', the organization won't be able to describe any criteria for a system other than cost, risk, technical requirements or stakeholder preference.

- **Collaboration with procurement specialists.** Procurement specialists can only work with the knowledge they have. Case studies have demonstrated that procurement departments do not have the skills to be able to express requirements to suppliers in user-centred terms.[21] This will not change without support and knowledge from user experience practitioners in other parts of the organization.

- **Documented quality and standards criteria.** General quality standards for the procurement of all digital workplace systems should be documented and agreed.

- **User testing.** Off-the-shelf products should undergo some form of user testing by potential end users before procurement takes place and the results of such a pilot should be one of the most important factors in the finalization of product choice.

A procurement checklist

As vendors rarely offer detailed information and reports on usability, DWG has developed a checklist to enable digital workplace teams and procurement departments to assess usability when choosing new applications.

It is a heuristic checklist (in other words, it enables you to assess the application against a list of commonly accepted principles). As such, it does not require deep technical knowledge or investigation (although someone from the digital workplace team who is familiar with usability should work with procurement in applying it).

It helps you ask the right questions when talking to vendors. And, because it can be difficult to get satisfactory answers during a product demonstration, the questions can also be used to structure visits you might make to organizations already using the application. The full checklist, with clarifying notes and examples, can be downloaded free from the DWG website.[22] It covers:

1. User authentication and support
When accessing the system, users should find it easy to authenticate. It should be clear from the interface how to get support:

- authentication should be simple and well supported in all use cases;

- it should be clear and easy to reset passwords;

- errors should be clearly communicated with user-friendly, platform-branded messaging;

- help lines and support information should be available and configurable to include a contact within the organization;

- forms and processes should be easy to follow and clearly indicate the requirements;

- data should auto-populate to support users;

- there should be validation of data to support correct entry and completion of information;

- help information should be focused and not overwhelming;

- technical staff within the organization should be able to monitor the system and intervene to support users.

2. Branding and appearance

The application should be easily configurable to conform with the organization's brand guidelines and standard behaviours:

- the application should have a consistent look and feel throughout;

- content should be clear and legible in all circumstances;

- key elements should be able to be modified to be consistent with the wider digital workplace.

3. Content compatibility

To be of most use to users, digital workplace services should integrate with and draw from a core content service so that content created within the application can be shared outside, or content from external sources accessed or presented within the application. This includes matching tone and style to prevent an incoherent experience. Creating compatibility will include these elements:

- user-friendly default text;

- system text that can be modified and additional text added;

- easily configurable terminology on all field labels, buttons, titles and navigation;

- multilingual interfaces and inputs supported;

- consistency of labelling on navigation, buttons, page titles, links and so on;

- export and import of compatible content.

4. Navigation and integration

Users should be able to navigate intuitively and without getting lost. Navigation should be similar to the experience of other parts of the digital workplace where appropriate:

- navigation into and out of the service should be intuitive and easy to operate;

- integrated services should match the base platform;

- interactive elements should be clear, consistent and easy to operate;

- submit and loading times should be clearly shown and not too long.

5. Mobile and accessible

Content and services should be accessible across a range of browsers and devices, including those designed to support users with impairments. There should be:

- graceful degradation on older and alternative browsers;

- style and content separation;

- keyboard navigability;

- colour contrast and legibility;

- text alternatives for media;

- responsive design for different screen sizes and devices.

VENDOR MANAGEMENT AND LOBBYING

Many software products become de facto standards in their industry and then changing them can be difficult. The vendor will have many existing installations to worry about and any change opens up the risk of annoying customers – customers, of course, who are the administrators of the systems rather than the end users. The vendor knows that inertia is on their side and that their customers are locked in. So why bother?

Organizations can, and should, be vocal over problems of user experience they identify with such products. What might seem to the vendor to be just a lack of polish, or to the owner of the system a minor internal annoyance, may actually be vastly increasing the total cost of ownership when time wasted is multiplied over a few hundred thousand employees. All that is needed is that the requirement (a description of what should be fixed) can be clearly codified and communicated to the vendor. However, the lines of communication are not always clear. Those able to properly describe the problem with a system in the digital workplace can be many people removed from the development team at a vendor on the other side of the planet. Some vendors are more open to suggestions than others; for example, Yammer has a customer network.

One solution is some form of industry-wide lobbying and gathering of such requirements broadly across many organizations, thereby adding a weight of authority. Another option is to crowdsource such complaints, using unofficial tools such as Get Satisfaction, Twitter or Google Plus to exert pressure for a particular change.

Over time, the experience of such tools will get better as development practices improve and vendors who are not serious about experience drop out of the market.

Chapter 12 Key Takeaways

- Most organizations are not looking at the user experience of the whole digital workplace, although work may be happening at the level of individual services.

- The current experience of the digital workplace is usually one of fragmentation.

- Underlying issues of how the digital workplace is managed and developed need to be addressed.

- Organizations have varying levels of control over different areas of the digital workplace: the designed and the non-designed.

- Poor user experience in the digital workplace leads to multiple issues including reduced productivity and employee engagement.

- Gaps in experience left by digital workplace projects will be paid for with the time and frustration of the users. This is the concept of user experience debt.

- Organizations need to develop a clear set of standards for the digital workplace as part of an overarching governance structure.

- The tools of user experience that are usually applied to individual sites and services can be applied to the digital workplace as a whole.

- In the designed digital workplace, issues of user experience can be addressed using interventions such as a user experience strategy, user research and mental models.

- In the non-designed digital workplace, user experience issues can be mitigated using interventions such as universal navigation, branding, personalization and enterprise search.

Notes

1   Schadler, Ted (4 February 2013) 2013 Mobile workforce adoption trends. Forrester: http://www.vmware.com/files/pdf/Forrester_2013_Mobile_Workforce_Adoption_Trends_Feb2013.pdf [accessed 28.03.14].

2   Miller, Paul (2012) *The Digital Workplace: How technology is liberating work*. London: Digital Workplace Group: http://digitalworkplacebook.com [accessed 28.03.2014].

3   Based on the median Fortune 1000 business in their sample (36,000 employees and US$388,000 in sales per employee). The University of Texas at Austin (undated) Measuring the business impacts of effective data. Sponsored by Sybase: http://www.sybase.com/files/White_Papers/EffectiveDataStudyPt1-MeasuringtheBusinessImpactsofEffectiveData-WP.pdf [accessed 28.03.14].

4   IDC (2005) The hidden costs of information work. IDC white paper: http://www.scribd.com/doc/6138369/Whitepaper-IDC-Hidden-Costs-0405 [accessed 28.03.14].

5   Webster, Melissa (2012) Bridging the information worker productivity gap: New challenges and opportunities for IT. IDC white paper, sponsored by Adobe: http://wwwimages.adobe.com/www.adobe.com/content/dam/Adobe/en/products/acrobat/axi/pdfs/bridging-the-information-worker-productivity-gap.pdf [accessed 28.03.14].

6   Eisner, Denise (21 February 2012) Special series – Taming the intranet beast, Part Five. Systemscope: http://www.systemscope.com/business-operations/special-series-taming-the-intranet-beast-part-five [accessed 28.03.14].

7   All About UX (undated) User experience definitions. All About UX: http://www.allaboutux.org/ux-definitions [accessed 28.03.14].

8   Nielsen, Jakob and Norman, Don (undated) The definition of user experience. Nielsen Norman Group: http://www.nngroup.com/articles/definition-user-experience [accessed 28.03.14].

9   It should be noted that the user experience field is in constant evolution and definitions of its role and scope vary among practitioners. One particular point is whether the term itself is self-limiting as the entire design process is focused on many stakeholders and processes beyond that of the user. For our purposes here, user experience and employee experience can be deemed to be interchangeable terms.

10 Lanier, Jaron (2011) *You Are Not a Gadget: A manifesto*. London: Penguin.

11 harmon.ie survey as cited in: Tata Consultancy Services (2011) Towards an integrated digital workplace. Tata Consultancy Services white paper: http://www.tcs.com/SiteCollectionDocuments/White%20Papers/Consulting_Whitepaper_Toward-Integrated-Digital-Workplace_01_2012.pdf [accessed 28.03.14].

12 Huddle (undated) State of the enterprise information landscape study. Huddle: http://www.huddle.com/files/research-papers/Huddle_study_-_State_of_the_enterprise_information_landscape.pdf [accessed 28.03.14].

13 Tubb, Chris (2013) Digital Workplace User Experience: Designing for a flexible workforce. Digital Workplace Group: http://www.digitalworkplacegroup.com/resources/download-reports/digital-workplace-user-experience [accessed 28.03.14].

14 iOpener Institute for People and Performance (2012) Job fulfilment, not pay, retains Generation Y talent. http://www.iopenerinstitute.com/media/73185/iopener_institute_gen_y_report.pdf [accessed 28.03.14].

15   Cisco (2012) Cisco connected world technology report. http://www.cisco.com/en/US/netsol/ns1120/
      index.html [accessed 14.04.13].
16   For a definition of technical debt, see Wikipedia: http://en.wikipedia.org/wiki/Technical_debt
      [accessed 28.03.14].
17   Bynghall, Steve (2012) Strategy and Governance: A good practice guide. Digital Workplace Group:
      http://www.digitalworkplacegroup.com/resources/download-reports/strategy-governance [accessed
      28.03.14].
18   Meister, Jeanne (1 April 2013) Flexible Workspaces: Employee perk or business tool to recruit top
      talent? *Forbes*: http://www.forbes.com/sites/jeannemeister/2013/04/01/flexible-workspaces-another-
      workplace-perk-or-a-must-have-to-attract-top-talent [accessed 25.03.14].
19   As 13 above.
20   Jennifer Watts-Englert, et al. (2012) Back to the Future of Work: Informing corporate renewal. EPIC
      2012 Renewal proceedings, p. 125. http://epiconference.com/2012/sites/epiconference.com.2012/files/
      attachments/article/add/EPIC2012-Proceedings.pdf [accessed 28.03.14].
21   Markensten, Erik (2003) Procuring usable systems – an analysis of a commercial procurement project.
      In: Julie Jacko and Constantine Stephanidis, *Human–Computer Interaction: Theory and practice*, p. 544.
      Mahwah, NJ: Lawrence Erlbaum Associates, Inc.
22   Marsh, Elizabeth (2011) and Bloom, Louise (2014 update) Usability of third party applications –
      checklist. Digital Workplace Group: http://www.digitalworkplacegroup.com/resources/download-
      reports/free-report-usability-of-third-party-applications-checklist [28.03.14].

Chapter 13

Setting Up the Digital Workplace Programme

Introduction

'Digital' is at the heart of organizational transformation in our current era. While it is often the external digital initiatives that attract the most attention, many organizations are starting to realize the potential of digital transformation that begins from the inside out. This is the digital workplace programme and it is being used to fundamentally transform organizations and develop new ways of working.

However, many organizations are entering this transformational era with little or no idea of how it will impact on their operations, let alone a plan for how to leverage the potential benefits and avoid pitfalls. In a 2013 survey of major organizations,[1] DWG found that only 36 per cent had a formal digital workplace programme or function in place. This is especially problematic because, as Julie Lakha points out in DWG research on this topic, from which many insights in this chapter are drawn,[2] digital workplace programmes can be complex, with 'multiple moving parts' at any one time.

Pioneers of the digital workplace are already starting to ask – and in some organizations answer – key questions about how to go about putting the strategy in place: Who owns the digital workplace? How do you attract senior attention? Where should you start in planning a comprehensive digital workplace programme?

In this chapter we will explore the answers to these questions, building on what we have already learned about the nature of the digital workplace, the business case for investment, and the crucial importance of keeping people and their work modes firmly at its heart.

In brief, this chapter looks at:

- taking an integrated approach;

- defining the strategy;

- establishing the governance model;

- implementation;

- a checklist for setting up the programme.

Taking an Integrated Approach

As we've already seen, one of the fundamental issues digital workplace programmes set out to tackle is the fragmentation of employee experience across an array of digital services and devices. To achieve this requires not only that people are kept at the core of the digital workplace programme but also that an integrated approach to scoping and delivering the programme is adopted. This includes deciding what is in scope, getting the right stakeholders on board, gaining buy-in, and establishing digital workplace leadership.

In DWG's research into the importance of this integrated approach,[3] Steve Bynghall highlights that one of the reasons organizations such as Unilever and the United States Patent and Trademark Office (USPTO) have been successful in their digital workplace programmes is because they have taken a cross-functional and holistic view of their digital workplaces.

We have already seen, in Chapter 11, how the lack of strategic alignment of collaboration initiatives is leading to

---

DIGITAL WORKPLACE EXPERT VIEW

The people who work for an organization – be they employees, contractors or even suppliers – must be at the heart of digital workplace planning. They do not want to know, or care about, the fact that the digital task they are trying to achieve, or the information they need, are on different systems or are delivered through different interfaces.

When a passenger wants to travel from the middle of Texas to the heart of France they need the train stations, airport terminals, underground systems and taxi routes to be connected – they don't want to be left stranded with no ability to connect to the next stage. They don't want to be told that on a key road, they can only travel in a lorry and not a car. The passenger (or user) needs an integrated plan that effectively and even pleasantly takes them from one mode of transport to the next. And wouldn't it be good if one ticket took them from the very start to the very end of the journey?

A seamless, single-ticket experience in the digital workplace may still be an aspiration – but good integrated and holistic planning can take us a long way towards it.

Helen Day, Group Managing Director, DWG

failure. Gartner predicts that, by 2015, 80 per cent of social business efforts involving enterprise social networks will not achieve their intended benefits; this is due to 'inadequate leadership and an overemphasis on technology'.[4] Our own research has demonstrated that pitfalls such as these can be avoided by getting the scoping right from the outset and ensuring that the right people are on board.

GETTING THE RIGHT SCOPE

One of the first steps in setting up a digital workplace programme is to consider the scope of the programme. For organizations to realize the impressive benefits we looked at in Chapter 11 – from cost optimization to productivity enhancement to accelerating innovation – they must take what Steve Bynghall describes as 'a widescreen and holistic view' when scoping the digital workplace programme. This view acknowledges the deep impact the programme will have on how and where people work in all areas of the organization.

A prerequisite for achieving the types of highly positive outcome that organizations such as Unilever, IKEA, Cisco and Virgin Media are benefitting from is that the scope of the digital initiative must:

- extend to a wide enough proportion of the workforce to ensure that gains are significant and fully leverage the digital investment;

- fundamentally change the way employees work, so that the benefits realized are significant in terms of process improvement and associated behaviours are embedded;

- extend individual choice of where and how employees work in order to result in improved employee engagement and a reduction in travel.

Achieving an appropriate scope for the programme will be easier if the strategic alignment has been made clear from the outset. Another factor that can help in scoping is where digital workplace change is part of a wider strategic transformation programme that automatically attracts cross-organizational involvement at a senior level.

INVOLVING STAKEHOLDERS

This scoping activity should highlight the various areas of the organization that the digital workplace programme will touch and, as a result, help to define which stakeholders need to be involved. Core stakeholders in most programmes will include those responsible for:

- overall company strategy;

- technology;

- physical workplaces and buildings;

- people-related policies and communication;

- change management and training.

The exact make-up of the stakeholder group will vary depending on the emphasis of the programme. For example, if flexible working is a key focus, then HR and Real Estate are two areas where a significant impact can be expected; or, if the organization operates in a highly regulated industry, then involving the Legal, Compliance or Audit teams at the outset can prove critical to the programme. Senior leaders from the relevant functions will be key, both as stakeholders for specific aspects of the programme and to fulfil roles within the governance structure of the programme.

As part of DWG's research into this topic, we asked respondents which internal functions were involved as key stakeholders at either a strategic or an ownership level in their digital workplace programmes. The results are shown in Figure 13.1.

With 82 per cent of organizations involving their leadership functions as a key stakeholder, it suggests that digital initiatives are either likely to be broad in scope or have strategic level importance.

There is also a relatively high proportion of people-centred functions (66 per cent HR, 45 per cent Learning & Development) involved as stakeholders, suggesting that organizations are recognizing that digital working changes the ways in which people work, and therefore requires strategic input on training and HR policies.

Our survey shows a relatively low level of input from Real Estate functions. This is reflected in the descriptions that respondents give of their digital workplace initiatives, many of which focus on social networking, collaboration or communications platforms. While it is perhaps not expected that Real Estate functions would be stakeholders in these projects, it seems that opportunities to align digital and workspace strategies are being missed.

Another consideration when determining the key stakeholders is that it is likely there will already be a strategy and/or development plan in place for any focus areas that have been identified. Bringing these stakeholders into the inner circle is more likely to achieve

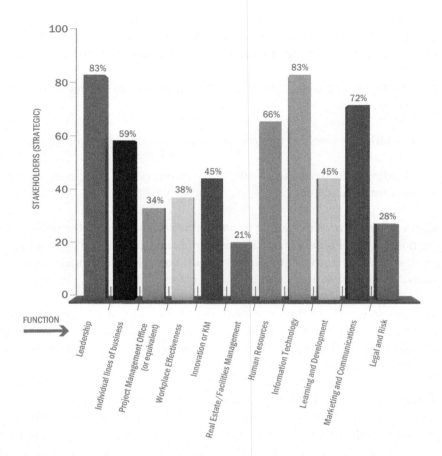

Figure 13.1   Functions represented as key stakeholders at a strategic or ownership level

buy-in and to smooth the process of realigning existing strategies and plans to the wider digital workplace programme. Finding the balance between direction and engagement is key, and clear ownership and accountability for delivering the programme essential.

TAKING A CROSS-FUNCTIONAL APPROACH TO DELIVERY

As well as ensuring that the right senior leaders are on board, the make-up of the digital workplace delivery team is also critical.

Technology projects have a high risk of failing to meet their original objectives. For example, a McKinsey study carried out with the University of Oxford found that large IT projects typically 'run 45 percent over budget and 7 percent over time, while delivering 56 percent less value than predicted.'[5]

The study found four key ways to avoid these problems, three of which touch upon taking a cross-functional approach to technology implementation, including:

- effectively managing projects with reference to business strategy and stakeholders;

- mastering technology and content with the right experts on board;

- building effective teams with a common vision, processes and a 'high performance culture'.

For a digital workplace project, having functional representation on the project delivery team from non-IT stakeholders is particularly important in order to avoid failure. This not only helps ensure that the right experts are on board but also makes it more likely the project will reference strategy and the needs of stakeholders.

In our survey, we also asked respondents which functions were represented as members of the project delivery team. Respondents could choose more than one answer. The results are presented in Figure 13.2, with representation at stakeholder level also shown for easy comparison:

The results show an under-representation of non-IT and Project Management roles in delivery teams across the board despite involvement at the stakeholder level. This is slightly less significant for the Marketing and Communications function, probably because they often have ownership of the intranet, a key channel within the digital workplace.

This disconnection between stakeholders and project team composition may be partly down to the scope of projects and the way in which governance is structured. For example, an HR stakeholder may sit on a digital steering committee and, for smaller projects, this involvement may be sufficient.

However, it may also be down to a traditional dominance of IT in project delivery, while other support functions, often tightly resourced, may be reluctant to release headcount.

In organizations with successful digital workplace initiatives, delivery teams are cross-functional. For example:[6]

- At Unilever, the delivery team is made up of people from HR, IT and Workplace Solutions.

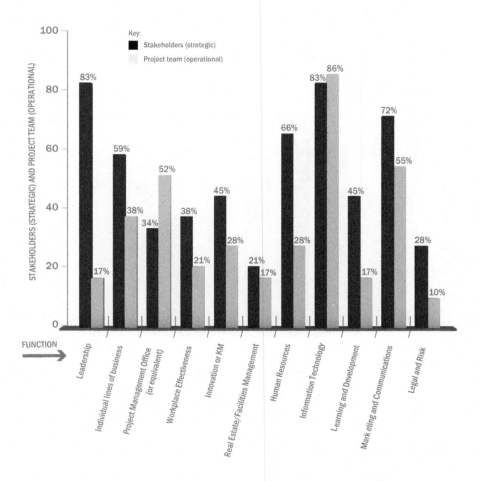

Figure 13.2   Functions represented as members of the project delivery team

- At Stora Enso, IT and Communications partnered to build their new digital workplace, and even established joint project managers.

- At Wiltshire Council there was close cooperation between a range of functions, including leadership.

- At USPTO, there has been enterprise-wide resourcing for their teleworking programme, with a central advisor helping to coordinate activity, and various cross-functional structures and committees.

Having cross-functional teams brings a range of clear benefits. In our survey we asked respondents about the outcomes they had experienced from working together. Respondents could select more than one option. The results are shown in Figure 13.3.

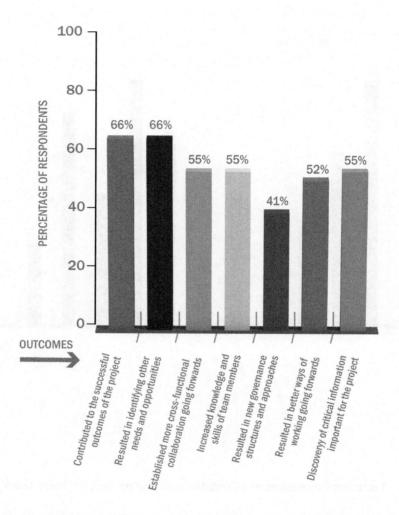

Figure 13.3   Successful outcomes from working across different functions

Apart from new governance structures, a majority of respondents indicated positive outcomes for all choices, demonstrating that the cross-functional approach was very important for that particular project.

Cross-functional delivery teams also help to build foundations for better project work through the upskilling of team members and the experience of working together. For example:

- Unilever recognizes the importance of cross-functional alignment and employs a change manager to support this.

- At USPTO, close collaboration is regarded as 'critical' for success.

- At Wiltshire Council, strong relationships have helped to shape future change programmes.

- At Stora Enso, working together between IT, Communications and HR led to a much better understanding of requirements.

Moreover, including cross-functional teams also means non-IT functions are likely to have a stronger sense of ownership, which may be important when new systems become 'business as usual'.

GAINING BUY-IN

Practitioners who have been operating in the intranet and digital workplace field for a number of years will remember the much-heralded arrival of the 'portal' and how this was going to be the 'one-stop shop' that would provide access to everything an employee would ever need. While successfully implemented in some organizations, others are still struggling to realize the benefits. There are potentially three key reasons why this solution was not embraced as quickly or widely as expected:

- In most instances it was seen as a 'technology project' and business awareness was not aided by the use of such terminology as 'portal'.

- Costs were substantial and making the business case for investment rather than the relatively cheaper option of carrying on with the content management systems and in-house developed technologies already in place was not an easy ask.

- To fully exploit the opportunity this technology offered would require a change in current processes and practices and, in most instances, business thinking had not sufficiently matured to embrace this new world.

So, how do you effectively secure buy-in to your digital workplace programme and avoid some of the pitfalls of previous technology projects? In DWG research into setting up a successful digital workplace programme,[7] Julie Lakha highlights four areas of focus:

Terminology – this is a business programme not an IT programme. Use terminology that resonates with both the business and IT; the term digital workplace appears to have gained traction due to its wide scope and the balance of technology and business focus it implies. It also encourages the audience to think outside of the normal constraints of a particular system such as the intranet.

Budget – if the digital workplace programme is recognized as an integral enabler for delivery of a wider business transformation programme, then securing an appropriate budget for the programme may not be an issue. If, however, the current economic climate, or experience of previous investment programmes that have not delivered expected ROI or benefits, have led to a lack of appetite for another multi-million dollar programme, consider identifying the significant budget already being outlaid on tools and applications, such as HR, document management and procurement systems. Bringing these together into a cohesive programme could enable more effective use of existing and/or planned investment as well as the opportunity to identify potential synergies.

Visualization – don't expect everyone to just 'get it' without some help. Visualization can be a powerful ally, whether in the form of a representation of the disparate systems and fragmented user experience you have today; visual mock-ups of what the new digital workplace could look like; use cases defining how an employee's working day could be different in the new world; or inspirational examples from other organizations that have successfully embarked on a digital workplace journey.

Hearts and minds – managing a diverse set of stakeholders and priorities takes effort and a perceived loss of ownership or autonomy can lead to resistance to change. In our experience at DWG, successful digital workplace programmes always show clear evidence of strong leadership, passionate evangelists and/ or change champions. Having a vision that is shared and understood by the stakeholders is also a critical success factor in achieving alignment.

DIGITAL WORKPLACE LEADERSHIP

A role that is emerging to lead digital transformation initiatives in major organizations is that of Chief Digital Officer (CDO). Gartner predicts that, although at the time of writing only 6 per cent of organizations have a CDO, this number is set to grow significantly, with 25 per cent of organizations expected to have a CDO by 2015.[8]

The creation of this new role appears to be a positive signal that digital is getting the recognition it deserves at a senior level. It also sends out a clear endorsement of digital transformation initiatives to employees. And while the CDO role straddles both external, customer-facing digital initiatives as well as the internal, operational side, some are seeing the transformation of the digital workplace as the foundation for the wider digital transformation of the company. For example, this 'inside-out' approach was taken by Indian paint manufacturer Asian Paints, who started with automating and streamlining internal processes with a view to providing a strong foundation for further growing the company.[9] We also see this principle at work in Spanish media company PRISA (see Chapter 14), where internal change is considered to be core to the overall digital transformation of the company.

However, the CDO role is certainly not a prerequisite for digital transformation and there is even some debate[10] as to whether its emergence implies the failure of the rest of the C-suite to grasp digital, or even a lack of understanding that such initiatives require a cross-organizational effort (rather than a potential new organizational silo). In practice, the digital transformation could be led by an existing senior executive, such as the CIO or CMO.

Either way, the success of the digital workplace programme is not reliant on the existence of the CDO role within the organization. What *is* key is meaningful senior sponsorship, in other words a senior leader who not only signs off budget but is also passionate about digital. In a recent McKinsey survey,[11] a larger share of respondents (up from 2012) reported that their companies' senior executives are now supporting and getting involved in digital initiatives. With 31 per cent of respondents stating that their CEOs personally sponsor such initiatives, McKinsey highlighted that this demonstrates not only the importance being placed on digital programmes for corporate performance, but also the challenges faced by many organizations in which the CEO is often the only executive with the mandate and ability to drive a programme that will impact the entire organization.

DIGITAL WORKPLACE TEAMS

Whether senior sponsorship comes from the CDO, CMO, CIO or even the CEO, digital teams are springing up in various shapes and sizes within organizations to help drive change. For example, PRISA created a central digital unit right at the start of their digital transformation programme.[12] As with the CDO role, the creation of a formal digital team sends a clear message of endorsement for digital initiatives.

As with intranet teams, the diverse skillset and ability to understand work modes and engage stakeholders across the organization are more important than which department the digital team sits within. In fact, Chris Tubb has pointed out that many of the skills of the intranet team are also core to the digital workplace team.[13] These include:

- stakeholder management;

- standards, controls and governance;

- managing complexity, providing structure and coherence;

- representing the user;

- balancing technology with purpose.

The most ambitious and forward-thinking intranet teams are leveraging this skillset to expand their remit across the digital workplace – a big but not impossible leap. The exact form the senior sponsorship and digital team take will vary depending on the culture and structure of the individual organization. What is important to note here is that the rapid emergence of both the CDO role and digital teams, in one form or another, are clear indicators of the level of opportunity and importance being attributed to digital initiatives.

Defining the Strategy

While many organizations are yet to establish a formal overarching vision and strategy, digital workplaces are already here: board members are using iPads to access meeting papers; corporate communications are on the intranet; collaboration capabilities are being implemented; self-service HR is becoming more prevalent; procurement is moving online; and the email in-box is the core focus of daily activity for many.

Each of these distinct areas will already have established its own strategies and roadmap and, in the absence of a single digital workplace vision or strategy, these will often have been developed in isolation and without reference to other employee applications. The business impacts of this can manifest in many ways, as we saw in Chapter 12, but at its simplest level could, for example, mean an employee having to maintain separate profiles in the employee directory, collaboration area and email service.

The single most important factor in defining the strategy is to ensure that it explicitly aligns and supports organizational (and, by consequence, business) objectives. This is the strategic alignment we looked at in Chapter 11. This could range from how the digital workplace will support financial targets, such as reducing cost:income ratios, through to improving employee engagement scores, or attracting and retaining key talent. Other key inputs will be: employee needs (which can be captured, for example, through employee surveys, one-to-one interviews and focus groups); external insight into best practice from other organizations; and an assessment of your digital landscape.

In setting out the strategy for the digital workplace, there are some important learnings we can draw from the development of intranets. DWG's intranet benchmarking[15] has shown that, in the period 2007 to 2013, developing and aligning strategy for the intranet continues to be an area of weakness:

- *a high proportion of companies do not have an up-to-date intranet strategy;*

- *where an up-to-date intranet strategy exists, it is often not explicitly aligned to corporate strategy;*

- *intranets without an up-to-date strategy have weaker senior sponsorship and 'governance'.*

In surveying intranet managers as part of our 2012 research into intranet strategy and governance,[16] we asked respondents to list some of the challenges they faced in mapping the future direction of their intranet. Key among these were disinterest of senior stakeholders, differences of opinion among stakeholders and resourcing issues. These factors probably contribute to the high incidence of organizations that do not yet have a formal intranet strategy. The digital workplace programme is certainly a different beast, already attracting much greater interest at a senior level; nevertheless, the learnings from intranets act to underline the importance of getting the digital workplace strategy right.

Alignment with organizational objectives, coupled with the potential breadth of services and applications that make up the employees' digital experience, mean that each organization's digital workplace strategy will be different. However, in DWG's experience, the need to be clear about defining the scope of the programme and areas of focus is evident, as this not only enables key stakeholders to be identified and projects aligned but also allows expectations to be managed at all levels within the organization.

> **DIGITAL WORKPLACE BENCHMARKING SNAPSHOT**
>
> When DWG benchmarks the digital workplaces of major organizations, one of the areas of capability and maturity it assesses is strategic alignment and management. Organizations at the early stages of their digital workplace journey often have IT management for individual services only and this may reflect local, departmental goals. As the digital workplace starts to mature, some of these services – such as the intranet, collaboration tools and core applications – may start to shift to a consolidated governance model and become more aligned to strategic goals. In a fully mature digital workplace environment all of these services come under a unified governance model, including all relevant externally hosted services, demonstrating total strategic alignment.[14]

When determining your approach, consider these areas:

1. The culture of the organization and how this may shape or influence your programme or, alternatively, how you will use this programme to change the culture of the organization; for example, by breaking down internal silos through the introduction of collaboration capabilities.

2. Research – investing in research can not only avoid expensive mistakes (such as building your flexible working approach around virtual access via a desktop when 70 per cent of your audience actually requires access via iPad/mobile/smartphones), but can also be used to support the thinking behind the strategy and approach when seeking wider buy-in.

3. Business needs – waiting 18+ months for the new digital workplace programme to deliver is unlikely to be acceptable. How can planned deliverables be realigned to the overall strategy and still deliver the required capability to the business? Are there any 'quick wins' or opportunities that can be exploited to demonstrate early value and create buy-in? When PwC set out to create a global social and collaboration network within the organization, they moved away from previous large-scale technology implementations towards a more agile approach. They delivered the project in a series of '90 day sprints' that enabled them to continually 'learn, listen and iterate'. See Appendix 1 for more information.

4. Budget – will the programme be funded centrally? Or is it being established on the basis of more effective utilization of existing and planned spend? If the latter, how will these previously assigned budgets be managed? In either scenario, some central funding will be essential to support training and change management activities as well as the development and implementation of global tools.

    The approach to funding can also be used to drive business behaviour. For example, if local units are charged back a percentage of the central costs of the programme (regardless of adoption), then approving additional expenditure to fund local developments that 'compete' with the central programme are likely to prove less attractive.

5. The impact of change – not only can change unsettle individuals and teams but it can also have a direct impact on the business as time is diverted into building knowledge, defining new processes and developing appropriate

skills and capabilities. If you have a Change Management/Transformation team within your organization, get them involved from the outset.

6.   Communication – if your programme is part of a wider transformation agenda, then this is likely to influence the style, approach and frequency of your communications. If not, consider whether you need a formal communications plan or whether this might set expectations that cannot be met (for example, possible delays to release dates). You may instead prefer to build awareness more informally, an approach adopted by Virgin Media (see Appendix 4).

7.   Third-party suppliers – organizations are increasingly working with external third parties to deliver products and services. How will this impact your approach? What requirements do you need to consider that will support the need to collaborate outside the firewall?

8.   Business value – how will you demonstrate business value? What measures and checkpoints will you put in place to ensure that this is delivered?

Finally, establish a detailed plan for the next three to six months and a high-level roadmap of proposed/planned deliverables for the next two to three years. This is the approach a major retailer took when planning its digital workplace journey and, while recognizing that the longer-term outlook would be less specific, the combination of a short- and long-term outlook enabled early deliverables to be demonstrated (maintaining leadership/stakeholder commitment) while also setting and managing expectations.

Establishing the Governance Model for a Digital Workplace

Organizations that have implemented new tools and technologies without the appropriate controls and guidance will have already felt the implications of poor governance (for example, duplicated and out-of-date content, volume of community sites spiralling out of control, inconsistent branding). Aside from the risk management implications and the fact that regaining control can be difficult, failure to have an effective governance model can also mean that opportunities to leverage investment and resources are not being maximized.

Most, if not all, organizations will already have some digital capability, and effective governance is one of the key enablers in moving from a collection of disparate and disconnected tools, services and applications to a holistic digital workplace that engages and supports employees in their day-to-day work. Governance models will evolve over time but are primarily influenced by existing cultures and organizational structures; for example, a strong 'control and command' from the centre or a federated approach with local countries having various degrees of autonomy. In all instances, effective governance requires strong ownership, a clear understanding of roles and responsibilities, and defined standards and policies.

ESTABLISHING OWNERSHIP

Moving to an overall governance model for the digital workplace will require change and, in some areas, a loss of autonomy. For those organizations that have already embarked on their digital workplace journey, senior-level ownership has proved to be a major factor in driving this change, ranging from enabling investment and resources to be unlocked, through to ensuring alignment of existing workstreams and the adoption of standards and policies. Ensuring that governance has senior attention will also assist in maintaining ongoing compliance with the standards and policies across the various businesses.

Strong Leadership at the United States Patent and Trademark Office (USPTO)

In Chapter 3 we looked at a powerful digital workplace example from USPTO.[17] Its leading telework programme has been strongly supported by senior management, who regard it as a means to deliver on USPTO's strategic goals. This means that enterprise-wide resourcing has been made available for the programme, rather than being confined to either a division or a function.

Danette Campbell, Director, Telework Program Office, USPTO explains:

Leadership that is committed to making telework part of its business strategy is key to building a successful telework program. Telework must be embraced by agency leaders and woven into the fabric of the strategic goals and mission of the agency.

At USPTO, support from agency leadership and close collaboration and cooperation between all of our business units has proved to be critical.

Successful telework programs are also designed and deployed by incorporating the necessary enterprise-wide resources. USPTO's program involves the Office of the

Chief Information Officer (OCIO), the Office of the Chief Administrative Officer (OCAO), and the individual business units in the agency, who have provided the necessary support to design and implement the program.[18]

DEFINING GOVERNANCE STRUCTURES

Digital workplace programmes can be complex, with multiple moving parts at any one time. It is therefore essential that everyone understands the role they will play and the routes for escalation if needed. When Cisco embarked on its digital workplace journey, considerable focus was placed on establishing a cohesive governance structure and on ensuring there were clearly defined and documented roles and responsibilities (see Appendix 2). Consider the following groups and their roles in your programme's governance model.

Digital Workplace Steering Group – the primary purpose of this group is to:

- set the overall direction for the programme, in line with organizational objectives;

- determine the strategic priorities and roadmap;

- prioritize investment and resolve prioritization conflicts;

- leverage investment and resources across the organization;

- ensure effective risk management;

- resolve any escalated issues.

Digital Workplace Working Group – this group takes direction from the digital board/ steering group and is responsible for:

- overall project management;

- driving core workstreams, for example, collaboration, content management, mobility;

- assessing current capabilities and identifying new tools and systems;

- management of roadmap and dependencies;

- specification and delivery of core components, for example, search, profile, single sign-on;

- collation and prioritization of business requirements;

- defining standards and policies;

- engaging with the implementation/delivery teams.

Depending on the scope of the programme, it may also be necessary to establish sub-groups at particular times to focus on a specific topic or area (for example, information architecture), the findings and outputs of which should be fed into the working group.

All of these groups should contain representation from the key functional stakeholders, such as Communications, HR, IT and Risk/Legal/Compliance, as well as key business lines within the organization. By involving these areas in the programme governance, buy-in becomes implicit and enables expertise to be leveraged; potential barriers or business risks to be identified; and new ways of working to develop.

If the number of stakeholders across all the different areas needing a voice is unmanageable, then consider establishing a core group and bringing additional stakeholders in and out of this group in line with development plans and priorities.

Each group should be underpinned by a formal 'terms of reference' or similar, which clearly articulates the role and mandate of each group and what is expected from each representative. In addition, governance decisions should be tracked and available to all stakeholders.

Cross-functional Alignment at Unilever

At Unilever, the Agile Working programme[19] – which is regarded as an example of a leading-edge workplace transformation initiative – is more than just a flexible work or office reorganization initiative. It seeks a deeper cultural change, removing artificial barriers around how work gets done, and enabling employees to choose when and where they work as long as the needs of the business are met.

Given its cross-functional remit, the group responsible for delivering the Agile Working programme is made up of people from the three relevant functions – HR, IT and Workplace Services.

Organizationally, all three functions are represented on the Agile Working steering committee, which consists of:

- Chief HR Officer (overall sponsor);

- Senior Vice President, HR, Global Categories and R&D;

- Chief Enterprise & Technology Officer (responsible for both IT and Workplace Services);

- Chief Information Officer.

The steering committee meets at least quarterly, with extraordinary meetings called in between if necessary. Some decisions on non-contentious issues may be taken by email or other communication channels.

Beneath the steering committee, a more operationally focused Working Programme Board also meets regularly, representing the workplace, practices and technology elements. Finally, there are the actual teams involved and a Change Manager.

Although HR has ultimate responsibility for delivery, in reality, the lead function varies from geographical area to area, depending on the priorities for that market. Jacobina Plummer, Global Agile Working Change & Communications Manager, explains:

> Part of my job is to try and educate geographies for the need for equal involvement across all three functions ... Getting cross-functional alignment is really important to successfully landing the culture change. For example, in Asia, IT has been very proactive but, as a result, Agile Working is perceived to be about technology. We are working really hard to ensure Workplace Services and HR engage too to ensure it isn't just seen as a tech programme by employees. [20]

ENSURING ONGOING GOVERNANCE

With the strategy in place and a governance framework established, the digital workplace programme is on a firm footing. The governance structures and ownership will evolve over time but it is critical that clear governance is in place that extends beyond the original project or programme. Ongoing governance will also need to consider:

- **Local governance.** Aside from the fact that to control everything from the centre would require an entire army of people, empowering local units to maintain elements of governance can also help to secure buy-in, create local ownership and provide formal channels through which any issues, concerns, learnings and best practice can be voiced. For example, Cisco's approach is to establish a Sustainable Management Team (SMT) for each area or neighbourhood, and these teams provide local governance in line with the central policy, enabling any issues or specific needs to be addressed locally. This, of course, relies upon there being effective standards and policies in place to guide as appropriate.

- **Standards and policies.** In Chapter 12 we looked at the importance of standards for the digital workplace. These standards, and how they are enforced, are an important factor in its ongoing governance. Alongside them sit the policies that underpin the employees' use and understanding of the digital workplace (see 'Organizational readiness' in Chapter 10). For example, a cornerstone of Cisco's Connected Workplace Strategy was to develop a 'Global Space Policy', which would underpin key programme deliverables, such as a move from individual occupancy to group assignment. In addition to the Cisco example of considering whether new policies are needed, identify any existing policies that could benefit or impact the programme (for example, a flexible working policy) and leverage/petition for change as appropriate.

- **Training.** No matter how intuitive you are able to make the digital workplace, employees need appropriate training to maximize their use of it, both on how to use the tools and how to adhere to policies. Training offered via e-learning or classroom-based methods will be just one aspect of a wider programme of activities aimed to help employees adjust to the new ways of working (see 'Implementation' below). For example, Intel created a Digital IQ training programme to help employees leverage social tools to increase innovation, communication and collaboration;[21] its online university comprises more than 60 courses available to employees, who can select a unique certification path based on their role.

- **Risk and compliance.** Involving Legal, Audit and Compliance at an early stage will help to ensure that risks are mitigated and compliance enabled. When reviewing the risks of your organization's digital workplace consider the following:[22]
  - Are your content standards, policies and processes minimizing the business risks from inaccurate information?
  - Are your third-party vendors doing all they can to protect the security of your information and do you have the necessary agreements in place to protect your organization and your customers in the event of a breach?

- Are your mobile devices offering enterprise-class security to protect your intellectual property and internal systems?
- Are your application development standards ensuring that you comply with accessibility, security and privacy requirements?
- Does your organization have the necessary mechanisms and forums in place to identify and manage risk appropriately?

Implementation

Digital workplace programmes mean organizational change. A structured implementation approach is therefore essential if expected business value is to be achieved. The formation of a cross-functional delivery team, as we saw earlier in this chapter, will be a key part of this approach.

CONTROLLED ROLL-OUT

Another key aspect is that digital workplace programmes will inherently consist of multiple workstreams and deliverables of increasing complexity. As organizations become more agile, the heady days of 'big bang' implementations seem to have gone, with proof of concepts, pilots, quarterly release cycles and phased roll-outs taking centre stage instead. This not only minimizes the risk of business disruption by enabling technologies and functionality to be tested in a contained environment, but can also build confidence within the project team and enable feedback to be incorporated into the design process.

IMPORTANCE OF CHANGE MANAGEMENT

Since the workplace will change and new ways of working will be required, effective change management is crucial. A significant level of focus on supporting employees throughout the transition (with, for example, classroom-based training, floor-walkers, on-demand support and genius bars) is evident and has proved a critical factor in driving adoption.

Both PwC and Virgin Media have made powerful use of advocates in embedding their digital workplace programmes (see Appendices 1 and 4). At PwC, a critical factor in the success of the global social and collaboration platform 'Spark' has been the setting up of a network of more than 1,000 advocates who have championed it and shown colleagues how to use it.

Larger organizations are likely to already have established change management processes and supporting expertise in place but, if not, there are a number of models available, such as Lewin's Change Management Model, the McKinsey 7-S Model and Kotter's 8-Step Change Model.[23]

Organizations with successful digital workplace programmes take change management very seriously and demonstrate just how critical it is to bringing employees on board with new ways of working.

Unilever Embed Agile Working in the Organization

At Unilever, there has been a series of change management activities to help embed the Agile Working programme throughout the organization. These include:

- twenty-five large-scale 'Agile Activation' events in different offices, reaching more than 10,000 employees;

- two hundred Agile Working training drop-in sessions;

- a global series of workshops conducted with HR business partners to help support them as facilitators for Agile Working;

- the launch of a global branded engagement campaign called 'Agile Working Unlimited', customized for local needs and including translated material; the emphasis of the campaign is on how the programme 'removes limits' across various different facets of work and life;

- a toolkit for facilitators from HR, IT and Workplace Solutions;

- the creation of an annual Agile Working report, including a contribution from the CEO, which focuses on progress, personal stories and wider business benefits, such as supporting business continuity during Hurricane Sandy and the 2012 London Olympics.[24]

Communication and Training at United States Patent and Trademark Office (USPTO)

At USPTO, to help embed new ways of working, the Telework Program has been backed up by communication efforts. Danette Campbell explains that a key element of the change management effort involves 'demonstrating the quantitative benefits of telework to include a positive impact on real estate, trademark and patent production, the environment, employee attrition, and transit subsidies.'[25]

There is also support and training. USPTO has a dedicated intranet site, which includes USPTO's telework policy, business unit guidelines, information about ergonomics for the home office, FAQs, organizational and IT points of contact for telework issues, asset management, telework case studies, research and media coverage about telework.

Training is also provided for both participants and managers. For example, within the Trademarks division, the focus of one of the annual management conferences was how to manage virtual teams.

Having a dedicated (and full-time) Senior Advisor for Telework has also been important. Danette Campbell has been tasked with overseeing the whole enterprise-wide programme, and is the primary point of contact for all issues. She also acts as an expert resource and prepares any necessary reporting.

Checklist for Setting Up a Digital Workplace Programme

We have covered a whole range of aspects related to setting up a digital workplace programme. You can download the following checklist from the DWG website:[26]

- identify the senior sponsor;

- define the scope of the digital workplace and what tools, services and applications are included;

- identify the key stakeholders across the functions, for example, Corporate Communications, HR, IT, Corporate Real Estate, Risk and Legal;

- identify the current business and system stakeholders for the tools and services that are defined in the scope and review existing roadmaps to identify any opportunities for quick wins;

- establish digital workplace steering and working groups, and define, agree and publish terms of reference that clearly articulate the role and mandate of the respective groups as well as individual roles and responsibilities;

- undertake internal and external research to capture employee, vendor/supplier and external insights that can be used to inform the strategy and approach;

- define the vision, strategy and high-level roadmap and secure executive sign-off;

- establish Key Performance Indicators (KPIs) and metrics;

- engage the Change Management team;

- review current strategies and roadmaps, and work with business and system owners to ensure realignment to overarching strategy;

- socialize the concept and begin capturing requirements for core functionality/components, for example, single sign-on, mobility, content management;

- review existing policies and define governance and design standards that cover the wider digital workplace, for example, intranet, collaboration, services, applications;

- determine the approach for roll-out of services and applications;

- establish delivery teams and local buy-in for implementation;

- liaise with Procurement to establish process and checkpoints for new applications and services being sourced.

Chapter 13 Key Takeaways

- Digital workplace programmes need to take a cross-functional and holistic view when defining scope.

- Cross-functional stakeholder groups and delivery teams are essential.

- New opportunities are emerging around digital workplace leadership roles and teams.

- Gaining buy-in and bringing stakeholders with you means, among other things, speaking their language and articulating a compelling vision.

- Define an overarching strategy for the digital workplace that aligns with business objectives and existing projects and programmes.

- Effective governance is a key enabler in moving from a collection of disparate and disconnected tools, services and applications to a holistic digital workplace that engages and supports employees in their day-to-day work.

- Senior-level ownership and clear roles and responsibilities within the governance structure are major factors in successful digital workplace programmes.

- Implementation is punctuated by proof of concepts, pilots, quarterly release cycles and phased, rather than 'big bang' roll-outs.

- The digital workplace programme will fundamentally change how people work; do not underestimate the change management effort required to help embed these new ways of working.

Notes

1   Digital Workplace Group (2013) DWG member forum survey (material confidential to DWG members).
2   Lakha, Julie (2013) Setting Up the Digital Workplace Programme. Digital Workplace Group: http://www.digitalworkplacegroup.com/resources/download-reports/setting-up-digital-workplace-programme [accessed 27.03.14].
3   Bynghall, Steve (2013) Digital Workplace Fundamentals: The integrated approach. Digital Workplace Group: http://www.digitalworkplacegroup.com/resources/download-reports/digital-workplace-fundamentals [accessed 28.03.14].
4   Gartner (29 January 2013) Gartner says 80 percent of social business efforts will not achieve intended benefits through 2015. Gartner Newsroom: http://www.gartner.com/newsroom/id/2319215 [accessed 28.03.14].
5   Bloch, Michael, et al. (2012) Delivering large-scale IT projects on time, on budget, and on value. McKinsey & Company Insights and Publications: http://www.mckinsey.com/insights/business_technology/delivering_large-scale_it_projects_on_time_on_budget_and_on_value [accessed28.03.14].
6   As 3 above.
7   As 2 above.
8   Hein, Rich (19 November 2013) Why the Chief Digital Officer role is on the rise. CIO.com: http://www.cio.com/article/743421/Why_the_Chief_Digital_Officer_Role_Is_on_the_Rise?page=1&taxonomyId=3123 [accessed 28.03.14].
9   Capgemini (1 February 2012) Digital transformation review – edition 2. Capgemini: http://www.uk.capgemini.com/digital-transformation-review-edition-2 [accessed 28.03.14].
10   Friedlein, Ashley (21 March 2013) Why a Chief Digital Officer is a bad idea. EConsultancy: http://econsultancy.com/uk/blog/62387-why-a-chief-digital-officer-is-a-bad-idea [accessed 28.03.14].
11   McKinsey & Company (2013) Bullish on Digital: McKinsey global survey results. McKinsey & Company Insights and Publications: http://www.mckinsey.com/insights/business_technology/bullish_on_digital_mckinsey_global_survey_results [accessed 28.03.14].
12   Capgemini and MIT Center for Digital Business (2011) Digital Transformation: A roadmap for billion dollar organizations. Capgemini/MITSloan Management: http://www.capgemini.com/resource-file-access/resource/pdf/Digital_Transformation__A_Road-Map_for_Billion-Dollar_Organizations.pdf [accessed 28.03.14].

13   Tubb, Chris (17 December 2013) Should the intranet team manage the digital workplace? Digital Workplace Group: http://www.ibforum.com/2013/12/17/should-the-intranet-team-manage-the-digital-workplace [accessed 25.03.14].

14   Learn more about Digital Workplace benchmarking at: http://www.digitalworkplacegroup.com/membership/overview/benchmarking-evaluations [accessed 28.03.14].

15   As 14 above.

16   Bynghall, Steve (2012) Strategy and Governance: A good practice guide. Digital Workplace Group: http://www.digitalworkplacegroup.com/resources/download-reports/strategy-governance [accessed 28.03.14].

17   See full USPTO case study in: Bynghall, Steve (2013) Digital Workplace Fundamentals: The integrated approach. Digital Workplace Group: http://www.digitalworkplacegroup.com/resources/download-reports/digital-workplace-fundamentals [accessed 25.03.14].

18   As 17 above.

19   See full Unilever case study in: Bynghall, Steve (2013) Digital Workplace Fundamentals: The integrated approach. Digital Workplace Group: http://www.digitalworkplacegroup.com/resources/download-reports/digital-workplace-fundamentals [accessed 28.03. 14].

20   As 19 above.

21   Meister, Jane C. and Willyerd, Karie (2010) Intel's social media training. *Harvard Business Review* blog network: http://blogs.hbr.org/2010/02/intels-social-media-employee-t [accessed 28.03.14].

22   Marr, Andrew (2011) Managing Risk in the Digital Workplace. Digital Workplace Group: http://www.digitalworkplacegroup.com/resources/download-reports/managing-risk-in-the-digital-workplace [accessed 28.03.14].

23   Normandin, Bree (28 August 2012) Three types of change management models. Quickbase.intuit.com: http://quickbase.intuit.com/blog/2012/08/28/three-types-of-change-management-models [accessed 28.03.14].

24   As 19 above.

25   As 17 above.

26   Lakha, Julie (2013) Setting up a digital workplace checklist. Digital Workplace Group: http://www.digitalworkplacegroup.com/resources/download-reports/setting-up-a-digital-workplace-checklist [accessed 28.03.14].

Chapter 14

Measuring Progress and Performance

Peter Drucker's famous quote is still ringing in our ears: 'If you can't measure it, you can't manage it.' Yet intranets have traditionally suffered from a lack of measurement: even where appropriate tools are in place, metrics are gathered and analysed inconsistently, and do not always help form the basis for decisions. Chris Tubb, in DWG's game-changing research on intranet metrics and measurement,[1] did not pull his punches:

> It seems that a decade-and-a-half after the widespread adoption of web technologies for use internally within organizations, many intranet teams are driving cars with fogged windscreens, no speedometers and no petrol gauges. Almost more alarmingly, when they do have these essential items, they often drive recklessly and with no regard to them.

For intranets, the tools, practices and principles that enable meaningful metrics are often only considered as an afterthought. For the digital workplace, with senior management backing and investment prerequisites to make the programme happen, digital workplace managers will need to put measurement in place from the start to demonstrate that it is fulfilling the expected benefits.

It is to be hoped that the high profile of digital workplace programmes means that they do not suffer the same problems around metrics and measurements as intranets. However, the reporting around internal digital tools and services is still generally immature relative to, for example, CRM data used for the management of external clients. Recognition that investment in the digital workplace is core to the wider transformation of the organization should help drive this maturity.

At the outset of the programme, it is important to identify appropriate measurement tools within the organization, or make the case for investment in them. Consideration should be given to how reliable data can be sourced and also how metrics will be captured as users move across different services and devices within the digital workplace.

Establishing key success criteria and a measurement framework against which you regularly report is essential and failure to give appropriate attention to this can lead to stakeholder disengagement and either a lack of investment or withdrawal of funding. In DWG research[2] on setting up the digital workplace programme, Julie Lakha outlined the areas that measurement should cover and we build on them in this chapter:

- performance against organizational objectives;

- overall programme progress;

- individual project status.

In reality, making direct links from the digital workplace programme to financial value or increased productivity can be complex and hard to prove. A holistic set of measures will be required that include quantitative and qualitative inputs to demonstrate how the programme is supporting organizational goals and realizing value.

Alongside the collation and analysis of appropriate metrics, organizations need to continually evolve the digital workplace programme as new needs become apparent and innovative technologies available. An iterative approach will be the lifeblood of the programme as it matures and becomes embedded into daily work processes.

In this section we look at:

- defining appropriate KPIs for the programme;

- tracking overall programme progress;

- monitoring individual project status;

- the power of success stories;

- iteration and ongoing research.

Defining Appropriate Key Performance Indicators (KPIs) for the Programme

In Chapter 11 we looked at the importance of strategic business alignment. The work done early on to align the digital workplace programme to business objectives will inform the critical success factors that drive the programme strategy. It will also highlight the sought-after benefit areas (for example, increased revenue, cost optimization, accelerating

DIGITAL WORKPLACE EXPERT VIEW

While organizations spend significant amounts of time and money analysing online behaviour and reporting on the performance of their external website against agreed targets and objectives, the data available around internal-facing services and tools is often basic and inadequate, with objectives poorly defined. This can lead to a lack of interest from senior management as well as perceived lack of value and increasingly subjective decision-making.

Yet employees are one of an organization's most valuable assets, and those organizations that recognize this and are applying the same rigour to defining success criteria and measurement disciplines internally, are more likely to deliver business value from their digital programmes and to gain competitive advantage through the improved efficiency and effectiveness of their workforce.

Julie Lakha, Head of Consulting at DWG

innovation), the basis of which can be used to support the development of KPIs and metrics to measure progress and performance. Without this insight into what is strategically important to the organization, and the desired benefits, it is impossible to select relevant KPIs, let alone scope the programme appropriately.

The benefit areas we looked at during the business case, with a range of associated KPIs and metrics that may be relevant, are shown in Table 14.1.

Although only certain metrics will be suitable as KPIs, nevertheless a range of metrics will be needed to help provide insight. For example, in measuring productivity (a complex proposition), metrics comparing the number of hours worked by office versus remote workers, or before and after implementation, time accessing VPN (for remote workers), number of calls per hour, or user satisfaction with new ways of working may be deemed contributing factors. As with all metrics, these need to be viewed as part of a wider picture to be fully understood and to ensure that signals of effort made or quantity produced are not mistaken for actual outcomes or quality of delivery.

For example, in her book *Sidetracked*, Harvard's Francesca Gino highlights the mistake of tracking a particular metric in isolation and without reference to related measures.[3] In the example she presents, a goal is set for a sales team, which is apparently met, however deeper investigation shows that in order to meet the target, staff were selling goods to friends at the end of the month, which were subsequently returned at the beginning of the following month. It was not until improving sales figures were cross-referenced with soaring return rates that the real situation became apparent. It's a sobering reminder of the need to ensure that targets and incentives drive the right behaviours, and to monitor a range of metrics that show the real results (our human tendency is naturally to see what we want to see!).

Table 14.1   Example KPIs and metrics linked to benefits emanating from digital workplace initiatives

Benefit areas	Example KPIs and metrics
Cost optimization	• Technology costs. • Development and support costs. • Employee:desk ratio. • Real estate infrastructure costs. • Business mileage.
People and productivity	• Adoption rates for tools and services. • Productivity (self-assessed or calculated). • Time to retrieve information. • Ability to access internal expertise. • Integrated processes that support business transformation. • Number of customer calls answered. • Number of hours worked. • Number of customer complaints/praise. • Level of absenteeism and turnover. • Attraction of new employees. • Employee engagement scores.
Business continuity	• System availability. • Number of employees able to work at home or in alternative location. • Productivity percentage during crisis situation.
Corporate social responsibility	• Number of business travel miles. • Greenhouse gas emissions. • Implementation and utilization of video-conferencing/telepresence rooms. • Energy costs.
Increased revenue	• Achievement of sales goals. • Bid:win ratio. • Improved capability/products/services. • Employee knowledge. • Customer satisfaction scores.
Accelerating innovation	• Number of ideas generated. • Number of ideas implemented. • Revenue generation from ideas. • Product development cycle speed.

HOLISTIC MEASUREMENT AT UNILEVER

Unilever has developed a targeted set of measurements for its Agile Workplace programme including:[4]

- Agile Certification for sites that have been transformed in accordance with company Agile design standards. This is a star rating system on a scale of 1 to 3. A site's score can go up or down from year to year, so there is an incentive for sites to continue investing and improving. For example, the certification includes an assessment of technology, including how current and effective it is – this is a powerful driver for ongoing IT investment.

- There is an Agile Working Global Scorecard that incorporates a range of measures around implementation and utilization of telepresence rooms; identification and implementation of new Agile Workplace sites; employee awareness of and training in Agile Working practices; and cost reductions from reduced travel and facility operating efficiencies.

Overall programme statistics are calculated including productivity (which is measured via self-reported engagement scores); improvements in space utilization; and reductions in environmental impact.

In this example, the KPIs span both the digital and physical workplaces, giving a very broad view. For each organization, the KPIs will be very specific to the exact scope of the programme and what this has set out to achieve.

Tracking Overall Programme Progress

In addition to measuring specific KPIs against organizational objectives, it is important to take a holistic view that shows overall performance and progress of the programme. This requires a baseline, a desired end-state and a framework against which progress can be assessed.

There may already be frameworks within your organization that can be used to assess progress (or which can be adapted), or perhaps external expertise providing an independent assessment will be of value. An update on progress against the roadmap (which supports the overall framework) should be included. Whichever approach is used, the year-on-year assessment of the digital workplace should help to drive the roadmap, to define goals and to highlight gaps.

In Chapter 10 we explored a model for assessing your organization's digital workplace at a high level. However, a more in-depth assessment, conducted independently, may also be required and this can have a significant impact on the programme. This is what we witness again and again at DWG: independent and detailed assessment can drastically improve stakeholder engagement, cutting through competing agendas and internal politics. As well as providing a baseline for performance at the start of the digital workplace journey, it can help to assess performance and drive the roadmap on an ongoing basis. In addition, an independent view can highlight weaknesses that need to be addressed (helping to justify investment and support the business case) and to validate successes.

DWG's framework[5] covers not only the digital workplace capabilities we looked at in Chapter 10 but also user experience and areas such as strategic management, the readiness of the organization and measurement. Such a framework can be used to assess a range of distinct digital workplace programmes across industries. As well as helping to define goals and identify gaps in the digital workplace roadmap, it can also demonstrate areas of weakness that may be due to the timing not being right to implement new services rather than the need to add new capabilities.

While KPIs will demonstrate performance against organizational goals, tracking overall programme progress in this way gives a broader insight into, among other things, how the programme is governed and strategy defined, as well as organizational readiness for digital transformation.

Monitoring Individual Project Status

Within the overarching digital workplace programme, individual projects and initiatives will need to be tracked with an appropriate set of metrics. These will contribute to the overall picture of performance. Depending on your approach this could be based around:

- specific capabilities and functionality (such as collaboration, document management or unified communications);

- roll-out to specific areas, functions or locations.

In both of these scenarios, try to establish a core set of metrics which you can supplement on a case-by-case basis as required for each implementation. This will enable easier identification of potential trends as well as highlight potential areas of underperformance and significant success, from which insights can be derived for wider benefit.

In *Measuring Intranets: A guide to intranet metrics and* measurement,[6] Chris Tubb explores a whole range of metrics for measuring intranet performance and many of these will be suitable for other digital workplace projects. In brief, these include:

- classic usage metrics, such as page views, visits, bounce rate and traffic sources;

- campaign metrics, such as conversion rate;

DIGITAL WORKPLACE BENCHMARKING SNAPSHOT

When assessing an organization's use of metrics, DWG looks for appropriate tools deployed to measure usage of different areas of the digital workplace. In organizations that perform well in this area, such tools are in place and analytics data are used to help drive strategy. KPIs are selected to track progress against strategic goals. All content and service owners actively monitor usage, drawing on the data to make improvements. There is also a process in place to regularly review the metrics and reporting.

- search metrics, including search failure or conversion rate and frequent search terms;

- sharing and interaction metrics, such as number of comments or replies, participation rate or time to question being answered;

- solicited opinion, which includes user satisfaction and survey data, use of the Net Promoter Score and content rating;

- business and management metrics, such as reach/coverage, adoption, cost avoided or time saved, and intranet benchmarking.

In addition, wider digital workplace metrics may include areas such as the number of processes or services delivered online, the quality of the data they deliver, availability across devices and the associated savings.

IMPLEMENTATION METRICS

As individual projects within the digital workplace programme are rolled out, a range of implementation metrics will provide valuable insights into performance:

- **Employee reach and coverage** – the size of the target group for the digital workplace change and the level of use by this group within a set timeframe after implementation. This is a useful implementation metric, especially when you can segment it by location or department and thereby see whether take-up is particularly high or low in specific areas of the organization.

- **Service availability and adoption** – these are basics for ensuring system availability and appropriate capacity for the audience, as well as levels and ease of adoption for new tools and services. For example, the number and types of calls to the helpdesk, or tickets raised, can indicate difficulties users are experiencing in getting started with the new tool.

- **User satisfaction** – for example, satisfaction with digital workplace tools and services, and perception of productivity levels as assessed by employees and managers. In addition, the Net Promoter Score, most often used to measure customer loyalty, might be used as a measure of employee loyalty to and pride in the company. Survey questions can help to highlight what is or isn't working.

- **Training availability and uptake** – if training associated with a new system or way of working is optional, extracting insights about its uptake can provide a useful indicator of awareness and potential resistance. Combined with insights drawn from employee reach and coverage, this can highlight areas where further change management efforts are needed.

- **Time saving, task effectiveness and efficiency** – reporting time savings following improvements to the digital workplace user experience, or perhaps the digitization of a process, is a powerful measure of impact. In addition, the effectiveness of completing the task (does the user give up before achieving it?) and efficiency of completing it (proportion of users succeeding on their first attempt) will provide valuable insights.

- **Cost and value** – if total costs for a system can be calculated, then useful metrics such as cost per employee can be derived.

SELECTING METRICS BY PROJECT

Earlier we looked at potential KPIs and metrics to demonstrate the benefits targeted by the digital workplace programme. Individual projects and initiatives will also need an appropriate set of metrics and KPIs that will contribute to the overall measurement programme, as illustrated in Table 14.2.

Table 14.2 Examples of KPIs and metrics for individual digital workplace projects

Type of programme	Benefit area(s)	Example KPIs and metrics
Introduction of collaboration tools	Cost optimization; people and productivity; accelerating innovation	• Number of communities created and used demonstrates levels of collaboration. • Number of complete profiles (i.e. including skills and expertise) demonstrates ability to locate expertise and share knowledge. • Reduction in number of emails sent with attachments within team demonstrates use of team space. • Reduction in time to serve customer due to support community.
Consolidation of content management systems	Cost optimization; user experience; people and productivity	• Reduced licence costs due to adoption of central platform. • Reduced development and support costs due to simplification of technology landscape. • Higher adoption rate due to greater simplicity of available tools. • Improved staff knowledge due to effective and consistent communications across centralized systems. • Reduction in duplicate processes and content as fragmentation is reduced.
Introduction of idea management platform	Accelerating innovation; increased revenue; people and productivity	• Cost to develop system plus ongoing operating costs. • Number of ideas generated demonstrates employee engagement. • Number of ideas generated that progress to project status demonstrates quality of ideas. • Revenue generated or costs saved due to ideas implemented.
Introduction of digital signature technology	Cost optimization; people and productivity; increased revenue; corporate social responsibility	• Reduced paper costs due to signatures processed online. • Reduced time to process contracts. • Improved conversion rate due to ease of signing up for customers.

LOCAL OWNERSHIP AND ADOPTION

In addition to measuring outcomes from specific deployments, demonstrating success or great use cases within a particular department or location can help to spur adoption and local ownership. At PwC, the team behind its social network and collaboration platform 'Spark' were not prescriptive about how and when each country or section should launch the new platform. This gave each territory ownership of the launch process and ultimately of their local 'Spark' site.

Paula Young, Global Head of Knowledge Management at PwC explains:

> *We didn't do a big global push. I allowed every territory to launch when they wanted and in any way that suited them, and that really worked. For example, in Switzerland there was a new Territory Senior Partner so they launched Spark when he started and it became part of the new Chairman's legacy.*

See Appendix 1 for the full PwC case study.

The Power of Success Stories

Dry facts drive most of us into a state of boredom. The smartest intranet and digital workplace professionals find engaging, as well as useful and relevant, ways to use metrics to demonstrate success – both inside and outside the organization. These may include success stories, well-designed dashboards, automatically generated reports or visualizations.

Success stories can demonstrate adoption in specific parts of the organization (helping to overcome resistance in others) and they can help to model the desired behaviours relating to the new tools. They can also become a formalized part of the toolset, directly incorporating the business case and financial value elements within it, as in the example of ConocoPhillips.

At ConocoPhillips, success stories help to demonstrate the value of communities of practice, which across the organization amount to hundreds of millions of dollars.[7] Each network can submit success stories via a form on the network homepage. These detail the business challenge, approach taken and an estimate of the cost savings or value generated. They are searchable across all networks, and the total value of all success stories is reported, to show the overall success of the programme. Each business division has a proportion of its annual bonus structure allocated according to its performance and behaviours relating to knowledge sharing.

The impact of such stories can be felt both internally, in terms of ongoing adoption of and investment in the programme, and externally, in terms of the organization's reputation. Walmart, for example, leverages testimonials from employees about their social network, mywalmart.com, so that hundreds of associates have offered testimonials that contribute to media, human resources, and internal and marketing communication. In fact, several stories have been turned into national advertising campaigns.[8]

When combined with rigorous measurement of the digital workplace programme, this external impact can be very significant. Cisco has taken KPIs a stage further and is a great example of how a robust set of KPIs and metrics can be used to demonstrate programme deliverables. Not only is performance reported internally but, as manufacturers of the technology utilized, the company also uses the data to report externally on its improved environmental impact as well as to promote its product capabilities. See Appendix 2 for further detail.

Iteration and Ongoing Research

While digital workplace professionals may fantasize about an ideal digital workplace that can be delivered in one fell swoop, the reality is that ongoing research, measurement, development and iteration will be required. In addition, as new tools and capabilities become available and desirable, there will be a need to continually innovate and experiment in order to ensure that the digital workplace experience stays relevant and up to date. And, of course, this is what keeps us excited about the digital workplace – the fact that it is never complete.

LEARN FROM USERS

Perhaps the best example of this is IBM's 18-year evolution of their early intranet into the IBM Workplace of today. Through a constant process of iteration, the team at IBM has developed its intranet from a 'consumption' to a 'participation' model, with people increasingly at the centre of the experience and impressive levels of adoption.[9] Both agility and openness to learning are modelled. In 2013, there was a reality check when levels of user satisfaction with the intranet were in decline, but this realization only served to inspire the team to analyse in detail data such as how users were using the homepage, leading to a redesign that massively simplified the user interface.[10]

The response here was ideal: the team was monitoring the signals from users about how they were using the tools (in this case, user satisfaction scores) and were able to mobilize to look across a whole array of metrics to diagnose the problem. This is a far cry from the 'fogged windscreens' we started the chapter with.

NO MORE BIG BANG LAUNCHES

As we noted in Chapter 13, the practice of the 'big bang' launch has fallen out of favour in recent years, mainly due to significant failures and dissatisfaction with this approach. Many intranet and digital workplace teams are replacing it with 'continuous improvement', with IKEA[11] and City of Malmo being good examples. In the latter case, a quarterly iteration cycle is reported to have brought a range of positive effects: it is easier to communicate about and promote the intranet; there is an implicit message that the intranet is already good (rather than needing to be rebuilt from scratch) and that constant improvements are being made; it bolsters the senior management perception that the intranet team is continually delivering value.[12]

ASK THE RIGHT QUESTIONS

We looked at user experience interventions in Chapter 12, and this is very much about ongoing research into the user's experience of working in the digital (and physical) workplace. This is illustrated well by the agile work programme that has been implemented at American Express.

The American Express 'Bluework' programme, which offers a range of new ways of working options to employees, has been widely reported and acclaimed for its success. Forbes wrote of its approach to ongoing feedback and development:[13]

> American Express regularly conducts a workplace survey 4–5 months following a BlueWork transformation to understand what had worked well and/or where there are challenges for the employees and leaders. Questions cover the work space, technology, change and communications, assignment of roles, support of leadership and satisfaction. Based on feedback, adjustments are made as needed to help support the business and employees. This type of feedback is crucial to making sure that flexible work policies remain productive and effective.

Evolving and Embedding Metrics

Over time, and as the programme matures, the KPIs and metrics that underpin measurement will also need to evolve. Certain measures may be relevant at the outset (for example, before and after implementation) but less effective or relevant over time. It is also important to look at how particular measures or scorecards may be built into the performance management structure. Linking particular measures to bonuses or performance records can send the message that digital tools and associated behaviours are critical to successful performance at an individual, team and departmental level.

HOLISTIC MEASUREMENT AT PRISA

PRISA is a Spanish language-based global media organization with businesses spanning television, news, radio and educational publishing. It has 11,000 employees based across 22 countries. In 2011, the organization recognized the imperative to digitally transform itself – both internally and externally.

The recognition that this transformation needed to start from within the organization has been a powerful driver for overhauling PRISA's digital workplace, according to Elena Sanchez Ramos, Chief of the Change Management Office, who is the lead for the programme:

> If we want to change the relationship with our clients, we have to change inside; we need to be able to talk in another language inside the company.

The programme also has strong links with the organization's journey to a single culture across its separate businesses, with the global intranet Toyoutome a key component of this cultural change. Over the last few years, PRISA Jobs, Campus PRISA (an online university), La Factoría de Experiencia (a knowledge platform in order to share the know-how and best practices from the finished projects portfolio of the company) and Mi Idea! (an open innovation platform) have all been delivered via the intranet to help drive changes in leadership, training, talent management and KM.

A culture of digital experimentation is also encouraged in the organization, to help make it more dynamic and competitive in the marketplace. CDOs within each of the separate businesses help to drive change and adoption, and also monitor performance and metrics.

At the outset of the digital transformation, PRISA developed a 'Transformation Index' to measure success internally and externally. Over time, the team has morphed its set of digital KPIs to fit with the way the programme has developed. In 2014, a key focus for the programme is on further developing the digital KPIs that are measured and reported across the organization. These KPIs are given additional credibility by the fact they are linked to managers' bonuses.

A 'Human Productivity Index' (HPI), which combines a range of quantitative and qualitative measures to understand the knowledge-based capital of PRISA, is also being developed. One axis tracks social participation and employee talent measured across the internal digital ecosystem (demonstrating the volume of knowledge transfer and social culture within the organization), while the second axis measures 'total productivity factors', such as total investment in training, innovation and human capital, plus the returns on these investments. Alberto González Pascual, Knowledge and Organizational Transformation Manager in the Change Management Office describes this as absolutely critical to the organization at this time:

> *It's about social capital and intellectual capital because all our workers are knowledge-workers. The incentive for the new behaviours we want to encourage is social prestige.*

PRISA operates in an industry that has already seen extensive disruption to its traditional model as a result of digitization. Its response has been to transform the organization from the inside out. The premise here is that the right digital skills, experience and culture within the organization, along with the ability to rigorously measure progress, fundamentally underpin the ability to transform and succeed externally.

It's a Journey

It has become a cliché to talk about 'the journey instead of the destination', but it is nevertheless true. It's also where we started this section of the book: understanding where your organization is on its digital workplace journey. And just as we arrive at the end of the book, it's timely to reflect that the journey is ongoing.

It's up to us, as digital workplace professionals, to remain restless and to some extent dissatisfied, because it is this that will keep us on our toes: always wanting to understand the new frontiers of technology and constantly relating what we learn back to how people really live and work. This is the 'digital work ethic' that Paul opened the book with, and it is up to us to model this new way of working; to be the evangelists in our organizations and beyond. Journalist and author Tom Standage rightly points out that: 'It is a sign of a medium's immaturity when one of the main topics of discussion is the medium itself.'[14] We talk about 'digital' a great deal right now – digital workplace, digital work ethic, CDOs and so on – but the 'digital' will become almost invisible as it becomes more embedded within every aspect of our daily lives. The opportunity for each of us is to evolve and reinvent ourselves, continually refreshing our expertise during this remarkable Digital Renaissance era.

Chapter 14 Key Takeaways

- Tracking KPIs against organizational goals, overall programme progress and individual project status are essential to the digital workplace programme.

- The work done early on to align the digital workplace programme to business objectives will highlight desired benefit areas and drive the selection of KPIs.

- A range of metrics should be measured to provide as broad and complete a view as possible.

- Tracking overall programme performance requires a clear baseline, a desired end-state and a framework against which progress can be assessed.

- Individual projects relating to specific capabilities and functionality, or roll-out to specific areas, functions and locations need to be measured.

- Using success stories – both inside and outside the organization – can be a powerful way of demonstrating outcomes and gaining ongoing engagement and support.

- The digital workplace is never 'finished' – a constant process of research and iteration is essential to keep it up to date and relevant.

- Metrics must evolve as the programme matures and develops.

Notes

1   Tubb, Chris (2012) Measuring Intranets: A guide to intranet metrics and measurement. Digital Workplace Group: http://www.digitalworkplacegroup.com/resources/download-reports/free-summary-measuring-intranets [accessed 28.03.14].

2   Lakha, Julie (2013) Setting Up the Digital Workplace Programme. Digital Workplace Group: http://www.digitalworkplacegroup.com/resources/download-reports/setting-up-digital-workplace-programme [accessed 27.03.14].

3   Gino, Francesca (2013) Sidetracked: Why our decisions get derailed, and how we can stick to the plan. Boston, MA: HarvardBusiness Review Press.

4   Pallant, Tim (8 November 2013) DWF Lab: Unilever's Agile Workplace programme. Digital Workplace Group Member News: http://members.ibforum.com/blogpost/439672/172266/DWF-Lab--Unilever-s-Agile-Workplace-Programme [accessed 28.03.14].

5   Digital Workplace Group (2014) Benchmarking Evaluations: Objective, expert feedback. http://www.digitalworkplacegroup.com/membership/overview/benchmarking-evaluations [accessed 28.03.14].

6   As 1 above.

7   Ranta, Dan (2013) ConocoPhillips: The power of connections. Slideshare: http://www.slideshare.net/SIKM/dan-ranta-power-of-connections-at-conocophillips [accessed 28.03.14].

8   Bliss, Surya (2014) Walmart's social network revolution. Melcrum case study: https://www.melcrum.com/research/harness-digital-technologies/walmarts-social-network-revolution [accessed 28.03.14].
9   Bynghall, Steve (23 October 2012) IBM's own social intranet journey. Digital Workplace Group: http://www.ibforum.com/2012/10/23/ibms-own-social-intranet-journey [accessed 28.03.14].
10   Freed, Ephraim (14 May 2013) Screenshots: IBM's massive intranet simplification. Digital Workplace Group: http://www.digitalworkplacegroup.com/2013/05/14/screenshot-ibm-intranet-simplificatio [accessed 28.03.14].
11   Lumbreras, Sonsoles (undated) How benchmarking helps IKEA keep its intranet in check. Simply Communicate case study: http://www.simply-communicate.com/case-studies/company-profile/how-benchmarking-helps-ikea-keep-its-intranet-check [accessed 28.03.14].
12   Bylund, Jesper (27 April 2012) Our intranet – six guiding principles, pt2. Jesperby.com: http://jesperby.com/2012/04/27/out-intranet-six-guiding-principles-pt2 [accessed 28.03.14].
13   Meister, Jeanne (1 April 2013) Flexible Workspaces: Employee perk or business tool to recruit top talent? *Forbes*: http://www.forbes.com/sites/jeannemeister/2013/04/01/flexible-workspaces-another-workplace-perk-or-a-must-have-to-attract-top-talent [accessed 28.03.14].
14   Standage, Tom (2013) *Writing on the wall: Social media – the first 2,000 years*. London: Bloomsbury Publishing.

Appendix I
Rolling Out Spark, PwC's Transformative Global Collaboration Platform

About Spark

PwC is a global accounting and consulting firm, offering services across 158 countries. The company employs over 180,000 people. As an intensive 'knowledge' business spread across so many locations, and with global clients expecting a joined-up seamless service, smooth and successful digital collaboration is critical.

Paula Young, Global Head of Knowledge at PwC, explains:

PwC is a people business and a knowledge business. Our central challenge is how do we get the best of PwC's knowledge and experience to our clients anywhere in the world every single day?

Because there are so many communities, groups and teams, which cut across different countries and divisions, PwC needed a single up-to-date digital platform to help facilitate effortless collaboration, often in real-time.

In 2010, a group of global partners, working with the global Knowledge function, helped to define an explicit goal:

To provide one common social networking and collaboration platform that accelerates our ability to connect with each other and collaborate together to create value for ourselves and for our clients.

The end result of setting this goal was the launch of 'Spark', a global social and collaboration network, which has received high levels of adoption, won awards and is gradually transforming the way PwC works.

Spark has many of the features associated with social platforms, including:

- social networking;

- blogs and discussion forums;

- the ability to easily configure sites at both group and individual levels;

- open, private and hidden groups;

- mobile access enabled;

- other social features such as gamification points;

- the ability to add content and web pages.

Approach

A distinguishing characteristic of the Spark project is that it is very different from previous large knowledge implementations carried out at PwC. These have tended to be complex and designed around extensive requirements. Paula Young likens the difference to that between an ornate cathedral and a bustling market bazaar:

> In the nineties, Knowledge Management was all about collection and putting things in the 'cathedral'. This meant beautifully architected databases and portals, but actually the 'congregation' was small. With enterprise social software we have the ability to create a 'bazaar' where people collaborate. The bazaar is a noisy and sometimes messy marketplace, but it's alive and dynamic with people exchanging and sharing knowledge.

Creating a 'bazaar' meant a new approach for PwC. From the outset, for ease and speed of implementation and future upgrades, it was decided to buy a platform rather than build one. The team also agreed they wanted to 'configure' rather than 'customize' and, in fact, Spark is more or less out-of-the-box apart from PwC branding. There are also no deep integrations with other PwC systems, although the product itself already integrates with Microsoft Office.

Another key approach was to take a more agile and iterative approach to launch. Paula Young explains:

We wanted to have a big vision but to get started as soon as we could rather than spending a year in requirements gathering. We also realized we didn't actually know how people would use the platform and that people might not know themselves. We wanted to experiment, as well as to listen, learn and iterate.

Timetable

To help keep up the momentum and adhere to an ambitious implementation schedule, the project was divided into a series of ten 90-day sprints, including four for embedding into PwC. It is notable that the actual speed of the 'design and test' phase was very quick with only one dedicated sprint, with the final launch scheduled in March 2012.

Moreover, the platform has never really come out of 'design phase', with new features regularly added, based on feedback and usage patterns. There is a monthly release, a quarterly release and a major annual release.

With a significant investment being made in Spark, and successful collaboration high on the agenda, the global team felt it was critical to handle the launch correctly so that decent adoption levels could be reached. Paula Young remarks:

We're a 180,000 people organization. The big focus when we launched was on getting critical mass, which is vital for these systems, but it's really hard to get people aware, engaged and bought in. We spent two sprints on this and then after that dug even deeper to drive value through specific uses and embedding behaviour.

Adoption

To facilitate adoption, some measures were put in place. The first of these was to call the initial roll-out phase a 'wave' rather than a 'pilot'. Not using the term pilot helped emphasize that Spark was here to stay. It also suggested the potential transformative nature of the platform and helped to drum up excitement. Calling it 'Wave One' also meant there was a sense of exclusivity. The result was that many parts of the business called up the global Knowledge function wanting to get involved. This outcome surprised Young:

I was inundated with people wanting to be part of Wave One. It sounds exciting. They didn't want to be part of Wave Two. It created a groundswell.

The team also knew it was important to select the right initial groups to work with so they could share success stories that would help drive adoption for future waves.

> We decided to go where the energy is. We didn't always go for the most strategic parts of PwC to launch Spark. Instead we went where there was a passionate partner who had a really clear idea of the value they thought they might get and wasn't going to fall at the first hurdle.

Young's team worked with around 20 different groups covering a variety of different countries and business groups like international tax and global sector offerings. In the set-up phase an individual workshop with each group was held, covering topics such as business goals and community management.

The end result was that some of these sites worked very well and are still going strong today, while others fared less well. With hindsight, Young feels she could have done more work with internal groups:

> On reflection I would have spent more time with the people who felt uncomfortable … It's taken some of the internal groups like Risk, IT and even Knowledge Management longer to get to grips with it than the actual business.

The Launch

Before Spark could be launched, it was necessary to ensure that any risk issues were mitigated; for example, establishing clear guidelines around protecting client confidentiality, meeting data protection regulations and ensuring IT security was robust. These were especially important as Spark is hosted, albeit within a private cloud. Appropriate terms and conditions for use were also established and the ability to report inappropriate content on each page was enabled, although in reality rarely clicked.

The team was not prescriptive about how and when each country or section could launch. This gave each country ownership of the launch process and, ultimately, of their local Spark sites.

> We didn't do a big global push. We allowed every territory to launch when they wanted and in any way that suited them and this really worked. For example, in Switzerland, there was a new Territory Senior Partner, so they launched Spark when he started and it became part of his new legacy.

There was also an array of imaginative launch activity varying from flash-mob dancing to branded air balloons.

Advocates

A critical factor in Spark's success has been the setting up of a network of over 1,000 advocates who have championed Spark and shown colleagues how to use it. Young recommends this approach to anyone considering launching a similar platform.

> *Before we launched, we spoke to about 15 other organizations. The ones that had been very successful with launching their social network all said that advocates were probably the biggest factor in their success.*
>
> *These are not your typical champions who get appointed. Advocates have to be taken from every level of an organization, from the receptionist to the global CEO. Advocates explain how they use the system in their part of the business. They are passionate, enthusiastic and do reverse mentoring.*

Advocates are supported by a community manager and their input is sought into the programme design. There are awards for advocates and they also participated in a promotional video for Spark.

Business Outcomes

One of the powers of Spark is that it has been created around business processes, including:

- creating global business proposals and tenders that often require significant interaction between teams – in one example they found that using Spark gave a 50 per cent faster to create time, with an 80 per cent reduction in issues around document version control;

- supporting global account teams who need to work more closely;

- helping to capture and disseminate market and sector insights;

- innovation;

- easing the onboarding process for new hires and trainees.

There are also a number of softer outcomes around people engagement; for example, facilitating conversations between senior management and staff in numerous countries.

One of the most common uses of Spark is for people to ask for help. Paula Young cites the example of a search for a specialist skill that had taken four days of fruitless searching, but when the request was posted on Spark, it bounced around the world and within 28 minutes the right person was located. Young declares: 'The speed and the reach are phenomenal.'

Becoming Established

Both adoption and feedback have been consistently good. Spark is even beginning to enter everyday vocabulary at PwC with the phrase 'I'll Spark you' established between colleagues, while some advocates dub themselves 'Sparklers'.

At the time of writing, the levels of adoption have been very encouraging.

- In the past 30 days approximately 50 per cent of the network have logged in, with 68 per cent logging in over the past 90 days.

- 69 per cent of employees recommend Spark to colleagues.

- There is an average of 300,000 page hits per day.

- The new platform has resulted in savings of approximately US$1 million due to closing global applications that are no longer needed.

Individual countries have also gained very high adoption (nearly 100 per cent in key markets such as the UK, Switzerland, Norway and Japan). In Japan and Norway, for example, it has in effect become their social intranet. Young points out:

> When we started we didn't compete with existing intranets and portals in different countries, but many countries are moving their intranet there.

Conclusion

PwC is already seeing a return on its investment in Spark. It has enjoyed high visibility, is clearly adding value and is achieving good adoption. But there is still a long way to go. Paula Young says:

> *The way people are working is changing in parts of the business, but we're still at the beginning of the journey and we're still learning.*

Appendix 2
The Journey Towards the Cisco Connected Workplace

Cisco is an American multinational corporation that designs, manufactures and sells networking equipment. At the end of 2012, Cisco employed 66,000 people. More than 15,000 people were based out of their American headquarters in San Jose, California and their global workforce works from more than 475 offices across 165 countries.

Talking with Alan McGinty, Senior Director, Global Workplace Solutions Group, it is perhaps not surprising that Cisco's journey to a more mobile, flexible, collaborative workforce had been taking place over a number of years. The evolving work styles of its employees had not gone unnoticed and, with a global property footprint of 22.5Mft^2 showing utilization at 50 per cent, it was clear that the promise of being able to work anytime from anywhere was gaining traction. Cisco was also conscious that the wider workforce was changing, as were expectations with the 'Net generation' being only five years away from entering the workplace.

Keen to understand more about the changing workforce and with an opportunity to realize potential real-estate savings, Cisco wanted a plan. It needed to support these new ways of working, providing the tools and technologies that would enable the workforce to work effectively as they became more mobile, and to convert a larger percentage of its footprint to a more flexible, collaborative way of working. The journey to a 'Cisco Connected Workplace' had begun.

Establishing the Programme

The first step was to gain a better understanding of the different roles and work patterns that were developing. The team settled on five core work styles:

1.   **Highly mobile:** travels extensively to customer and partner locations; frequently interacts with customers.

2.   **Campus mobile:** internally mobile; interacts cross-functionally in face-to-face scheduled meetings; often in leadership roles.

3.   **Remote/distant collaborator:** non-mobile employee who works frequently with remote colleagues; and frequently works from home.

4.   **Neighbourhood collaborator:** neighbourhood-based employee who is mobile within the group area; interacts with, coordinates and manages teams.

5.   **Workstation anchored:** desk-bound, non-mobile employee who performs highly focused individual work; some team interaction.

Recognizing that there would be a blend of styles across the organization, these were then mapped across key business lines and further analysed into mobile, remote and fixed work categories. The results were compelling – every business line mapped showed the fixed work category in the minority, and in the sales and services area this was as low as 5 per cent.

Culture was also an important factor, although as Cisco had become increasingly global, the majority of employees were already telecommuting at least once a week and working in a different location to their manager.[1] The 2010 *Cisco Connected World Report*[2] had some interesting insights:

- participants preferred jobs with workplace flexibility and remote access at a lower salary to less flexible jobs at a higher salary;

- nearly half of those with remote access worked up to three extra hours a day; a quarter worked four or more hours;

- two out of three employees (66 per cent) expected IT to allow the use of any device – personal or company issued – to access corporate data anywhere, at any time.

The scene was set and, with HR, IT and Finance fully engaged, Cisco set about developing a multi-year plan that would transform the employee workplace.

Strategy and Approach

Cisco's Connected Workplace strategy is based around four quadrants:

Effectiveness: Work performance and productivity.

Engagement: Employee engagement/retention.

Efficiency: Portfolio utilization and services improvement.

Environment: Reductions in energy demands and CO_2 emissions.

Combining the concepts of Policy + Technology + Environment, Cisco's approach was to develop a 'Global Space Policy', which would:

- support a move from individual occupancy to group assignment;

- integrate desk phones and mobiles to support mobility;

- introduce capabilities that encouraged collaboration (such as video telephony);

- develop a complete design solution set, which included performance and environmentally sustainable specifications for furniture systems, carpets, lighting, and so on.

A proof of concept was undertaken in San Jose and a detailed study on the people and financial benefits showed:

- Prefer environment +77 per cent

- Improved communications +82 per cent

- Workforce satisfaction +82 per cent

- Ease of finding a quiet space +62 per cent

- Ease of finding meeting room +80 per cent

- Reduction in space and capital expenditure equalled a ROI in under three years.

The overall programme was approved by senior management.

Communication

From the outset, Cisco established two principles:

- this was not a 'one size fits all' programme (as each group would have its own unique needs);

- the importance of effective change management in order to drive adoption.

Each building or project that rolled out would therefore have its own change management and communication plan, while wider awareness would be supported by regular company-wide articles on the overall programme and different benefits.

Governance

Recognizing the significant impacts of the programme on the property portfolio and the logistics of re-engineering existing buildings to meet Cisco's needs, there is a global team of 115 in-house professionals and an extended network of outsourced partners covering strategic planning through to real-estate transactions, facility management and product development.

Local sponsorship is secured for every new roll-out, and each specific area or neighbourhood has its own Sustainable Management Team (SMT). SMTs provide local governance in line with the Global Space Policy, enabling any issues or specific needs to be addressed locally, and are comprised of an Office Council of local business representatives and the Workplace Resources and HR teams that support them. The central team monitors and conducts pre- and post-implementation surveys to understand the net change to individuals and what needs to be changed, mitigated or developed.

Temperature checks with employees are captured through a number of different groups that meet (Employee Engagement, Employee Listening, Workplace of Choice), as well as regular assessments of interns and younger workers. Results of biannual work profile and client satisfaction surveys are also compared to past performance and any areas of change identified.

Implementation

In order to build momentum, initial roll-outs were targeted at groups within Cisco that were already interested in moving to a more collaborative environment.

With each implementation tailored to meet the unique needs of the respective group or business, historical data on building usage are used to populate a programme template and provide a breakdown and ratio of all space types. This is supported by a universal planning guide that defines high-level performance specifications and enables appropriate solutions to be identified.

As well as the logistical challenges of re-engineering its existing property portfolio, Cisco recognized the potential impact on its employees and the resistance that can occur with change. A comprehensive change management plan is developed for each roll-out and supports engagement with stakeholders and employees at all levels through, for example, facilitation of orientation sessions, workshops and focus groups.

Thorough Measurement at Cisco

Cisco has set itself a number of measures against which it actively reports and tracks the benefits delivered by the programme, including:

- floor-space utilization;

- employee effectiveness and engagement;

- client (employee) satisfaction scores;

- efficiency savings;

- environmental savings.

Not only are these measures used internally to demonstrate the benefits achieved through the programme but, as manufacturers of the technology utilized, specific references to performance achieved are also used to promote product capabilities and Cisco's environmental impact, such as in Cisco's 'Citizenship Report':[3]

> Typically, Cisco's meeting rooms are overbooked, while offices and cubicles remain vacant for up to 65 per cent of the time. To address these inefficiencies, Cisco Connected Workplace combines collaborative and networking technologies with an open floor plan and an emphasis on mobility, thereby reconciling productive working patterns with environmental responsibility.

> Cisco Connected Workplace increases workplace efficiency and has enabled a 40 per cent increase in employees assigned per 100,000ft² office space, which has a significant impact on the need for and cost of real estate, and benefits the environment by reducing new construction.

Reducing the number of devices in the workspace has the quantifiable benefit of reducing power consumption. In one building at Cisco, the Connected Workplace implementation resulted in 50 per cent fewer Ethernet ports, significantly fewer shared devices such as printers and copiers, a higher density of wireless access points, and the elimination of stationary personal appliances in the workspace.

Reflections

The programme is still being rolled out across Cisco, although McGinty believes that the two most important factors throughout have been early and effective senior-level communication and comprehensive change management activity:

There is always more change management activity you can do, such as communications, training, workshops, focus groups, floor-walkers etc., and always learnings with every additional project; the key is having the ability to course-correct as you go forward.

Notes

1   Cisco (February 2012) The Expanding Role of Mobility in the Workplace. A custom technology adoption profile commissioned by Cisco Systems: http://www.cisco.com/web/solutions/trends/ unified_workspace/docs/Expanding_Role_of_Mobility_in_the_Workplace.pdf [accessed 03.04.14].
2   Cisco (2010) The Cisco Connected World Report: Employee expectations, demands, and behavior – accessing networks, applications, and information anywhere, anytime, and with any device, October 2010. Cisco Systems Inc: http://newsroom.cisco.com/dlls/2010/ekits/ccwr_final.pdf [accessed 03.04.14].
3   Cisco (2008) 2008 Cisco Corporate Social Responsibility Report. Cisco Systems Inc: http://www.cisco. com/assets/csr/pdf/CSR_Report_2008.pdf [accessed 03.04.14].

Appendix 3
Creating a Better Place to Work: Microsoft's Workplace Advantage Programme

Microsoft's Workplace Advantage (WPA) programme is one of the world's most admired Agile Workplace programmes, helping to drive more efficient and enjoyable workplaces. It has been running for over a decade and has been responsible for transforming the firm's existing physical workspaces, as well as creating new environments for showcasing technology. Although the programme is primarily about transforming physical space, it is essential that this works in harmony with both technology and the way in which Microsoft employees need and want to work.

Origins

Although the programme started formally around 2003, its roots go back to the mid-nineties when Martha Clarkson, Global Workplace Strategist at Microsoft, was tasked with improving the work environments at the campus buildings in the Puget Sound area. The company was expanding significantly and was finding its space no longer fit-for-purpose. These days Clarkson is responsible for the global programme. She recalls:

> The campus was built on a belief that software programmers needed a closed door and quiet and privacy. There were maze-like corridors and it was hard to find people. All the walls were medium grey in our rainy climate. It was clear a lot of work needed to be done.

After various projects on the campus, the company decided to step back and research how it could make its work environment map more closely to business success. The 'war for talent' was at its height and Microsoft was now a far more diverse operation with a number of different teams. Clarkson explains:

> We had changed considerably so we said: 'We really need to understand who our internal clients are.' We spent a year and a half carrying out research. We

interviewed senior leaders and employees, conducted focus groups and surveys, and then selected key sites for time use observation.

During the research and also in designing output, the WPA team got input from HR and business leadership as well as other functions within the business.

Out of the research came four basic principles, which still form the basis of the programme. These are:

- enhance innovation and productivity;

- function-based workspaces;

- attention to human factors;

- showcase Microsoft brand and technologies.

Enhance Innovation and Productivity

From Clarkson's research and also her own experiences at Microsoft, it's clear that teams have very different working patterns. In particular, at Microsoft there tend to be two different types of work location in terms of activity and culture – one revolves around the engineering and programming side of things while the other facilitates sales and marketing. Even within these two main groups there are further diverse sets of employees with different working patterns. Enhancing innovation and productivity for employees means that one size doesn't fit all.

Martha Clarkson comments:

We want to be flexible, although we still have to be operationally smart, so you just can't have custom everything for everybody. Finding a balance is important. Ultimately we want to keep people creative and productive. It's not about real estate but about keeping the business successful.

The needs of different teams can vary greatly and space is increasingly being viewed in terms of 'neighbourhoods', which serve up to 16 people. The WPA programme allows each 'neighbourhood' to help define its own space and the appropriate integration of technology:

We are really trying to design in tandem with the group and let them create the protocols for the space as they want. When teams are enforcing their own ideas it

*just works much better. It's really important for groups to have that ownership
and to be able to run the space as they want to.*

Collaboration and Human Factors

The second principle around function-based workspace includes providing more
collaboration spaces. This partly reflects Microsoft's original office plans, which simply
didn't include enough collaborative space, and also the general movement to a more
collaborative work style. Agile programming is the software process that best embodies
the need for collaboration.

The third principle – human factors – is also important. This includes paying
attention to factors such as daylight, colour and materials, natural materials and also
providing a variety of experiences, which Clarkson views as crucial since 'people are
going to spend so much time in the workplace'.

Showcasing Technology and Brand

As a technology company with enabling collaborative technologies, such as Lync and
Windows Phones, it is inevitable that this is going to influence workplace design. Martha
Clarkson:

> *There's a huge sense of pride and community from employees in showcasing their
> work and we want to make sure we allow for this within groups. They're proud of
> what they do. For instance, someone from one area of the business might work with
> someone from another division such as Bing. Once they begin working together
> they immediately get a feeling of what each other does and what their culture is.*

External customers will also be visiting offices:

> *For the sales sites showcasing technology is critical because we're trying to sell our
> products. It's even more important as we move towards being a devices company.*

Programme Evolution

After the research phase, Microsoft undertook a number of different pilots in 2005.
Clarkson explains:

We went to six different business leaders who already understood that WPA would help support their aims and who were interested in making a transformation. We then took our learnings from these and went big scale.

As well as the pilots, a Workplace Lab was built which showcased a kit of parts for work environments, as well as how technology could be integrated.

After this, the team started to create new buildings from the ground up, including a tower near Seattle, which was all open space. By 2012, over 74 buildings worldwide had been included in the WPA, with 29 still in progress. A notable success has been the Schiphol building in Amsterdam, which has recorded a 40 per cent increase in employee satisfaction over a number of years.[1]

Approaches

At its heart the WPA progamme has focused on improving opportunities for employees. Clarkson says:

Microsoft is very entrepreneurial. So there's no mandated order based solely on cost reduction. First and foremost, we're working to keep the people in the business creative, productive and successful by providing more choice.

There are various other measures in place to support the programme. The Global Workplace Strategies team uses different metrics to show progress.

We did a baseline survey on various factors back in 2009 so we could then measure improvements and progress against that. Amongst other things, we look at levels of collaboration and employee satisfaction with the workplace.

Productivity is also recorded, with internal studies demonstrating that a 6 per cent increase can be achieved.

The team also collects inspirational use cases:

In terms of stories, we capture these on video. If a business leader can cite an example of something that's changed for the better on camera, then this is a powerful story for the next group that is going to undertake the transformation. We make sure these are fast, short and playful. They usually include some impactful sound bites.

Change management is vital, with the team hiring a change management consultant for each project and executing on 'WPA Readiness', a framework that allows agile deployment of the transformation. Clarkson believes this can 'really make or break the project for the end user'.

An Example of Change

Under the Workplace Advantage programme, each separate location may experience something different, but a relatively typical example is Microsoft's Singapore office.[2] Here:

- research was conducted, which suggested that 45 per cent of all employees were highly mobile;

- projected growth in headcount suggested that space would run out;

- there was an increased demand for space dedicated to customers.

To enable the change:

- unallocated seating was introduced;

- full mobility was introduced, both inside and outside the office, through the roll-out of more cloud-based solutions, collaborative tools such as Lync, and devices such as Windows Phones;

- desks were introduced with just one network cable required to enable an integrated monitor and hub, with wireless mouse and keyboard, helping to maximize desktop space;

- an increase in wireless connectivity capability (to handle increased traffic), Enterprise Voice and proximity printing (print to any printer) were enabled;

- new headsets were issued for all staff.

Of these technologies, Lync is possibly the most important. This has video conferencing as well as audio phone and text messaging capability, meaning most communication can be managed from one interface, both internally and externally.

To make things easier for employees to locate available desk space there is a solution which:

- shows the location of free workstations and bookable rooms in the building in real-time;

- allows employees to book space and rooms;

- allows employees to easily 'check-in' to a location and view where colleagues are.

Conclusion

Microsoft's Workplace Advantage continues to make a significant impact in the organization. Some approaches have been key in enabling this. These include:

- focusing on business benefits;

- recognizing the differences between various teams' needs;

- ensuring there is change management.

Perhaps most importantly, the programme has always been rooted in a solid understanding of user needs and working patterns, based on robust research. Martha Clarkson and her team also carry on learning from the programme and evolving it as necessary.

Notes

1   Microsoft (undated) Microsoft Workplace Advantage Program (WPA). Workplace Innovation Group: http://www.workplaceinnovationgroup.com/wp-content/uploads/2012/04/Microsoft-Workplace-Advantage-Program.pdf [accessed 03.0414].
2   Microsoft (September 2012) Transforming the Microsoft Workplace in Singapore. Microsoft IT showcase: http://www.microsoft.com/en-gb/download/details.aspx?id=34678 [accessed 03.04.14].

Appendix 4
Engaging Employees and Joining Up the Organization: The Virgin Media Digital Workplace

Virgin Media was the UK's first provider of all four broadband, TV, mobile phone and home phone services and is part of Liberty Group, the world's largest international cable company, which serves 25 million customers across 14 countries. It has around 20,000 employees.[1]

Virgin Media's digital workplace journey started with a desire to reduce employee travel and realize cost savings through reducing its property portfolio. What started out as a pilot in 2011 successfully moved on to a wider deployment to all employees in 2013 that empowers them to work in new flexible ways. Simple, intuitive tools that work across devices are enabling work to happen in any location.[2, 3]

Towards a New Way of Working

The initial impetus for the flexible working proposition came from Virgin Media's 'Leadership Edge' programme, an initiative that gives talented individuals the opportunity to excel. They had homed in on the amount of money the company was spending on travel between different sites (the property portfolio had expanded through merger activities) and had identified flexible working as a potential solution. With backing from senior management, a project team was established to pilot new technologies with the goal of enabling employees to work flexibly and find a better work–life balance, while also reducing operating costs.

The project team was cross-functional (a factor DWG research[4] has shown to be critical to success), with representation from IT, internal communications, finance, procurement and legal departments – as well as Virgin Media's external partner, Cisco. The impetus to work across organizational boundaries was also a key driver for the programme, which created new and effective ways to collaborate and break down traditional silos.

Establishing the Programme

A pilot was established in 2011, with a roll-out of the technology to 1,000 employees across the organization. A clear business case was made: 'We worked out that the one-off costs, such as internal labour, technology and OpEx costs, were overwhelmingly outweighed by the potential bottom-line impact in terms of productivity, engagement, travel and property savings,' according to the Adoption Lead for Collaboration Technologies, Leon Benjamin.[5] The pilot included:

- webcams and headsets to enable participants to benefit from the new audio and video options;

- an add-on (CUCIMOC) to their instant messenger service that turns it into a telephone calling device;

- audio and video conferencing (Webex) so that participants could avoid travel and instead conduct meetings from their desk or home office;

- an enterprise 2.0 platform (Cisco Quad) to make it easier to call colleagues, embed video links in messages and form online communities – all ways to help break down silos and create greater transparency in the organization.

Creating a Human-centred Digital Workplace

One of the key strengths of the Virgin Media Flexible Working programme was that it was focused on people right from the start. Instead of talking about technology features, the team focused on how the technology could help people get work done more easily, enabling simple, everyday tasks. Virgin Media's Director of IT Technical Services, Colin Miles, explains: 'The best way to transform a company is to create connections between people who previously did not interact. Cisco WebEx Social has become a virtual water cooler where people can meet and get to know each other outside of a particular task.'[6]

In addition, the approach to establishing adoption has been viral rather than company-mandated. Commenting on Virgin Media's 2012 Peer Award 'Technology for Performance' award, Leon Benjamin said: 'We shifted from a command and control model of thinking to a network-centric model of thinking where adoption happens in a viral "bottom up" fashion rather than "top down".'[7]

This approach to adoption also includes the idea of 'Super Connectors' – influential employees who have strong informal networks – to help influence wider employee attitudes to the tools.[8] These people were identified via analysis of email statistics internally, as well as LinkedIn, Twitter and Facebook externally.[9] They were then brought in to the pilot and trained up in how to run a community and recruit members. This approach is creating a 'ripple effect' of adoption through the company and is a response to one of the main challenges the pilot faced of getting additional employees onboard with the new tools.

Adoption has also been supported by the recognition that there is a range of ages and technical abilities among Virgin Media's employee population, and that 'one size does not fit all' in terms of training. As well as a traditional help desk, a variety of training options is offered including one-to-one training, self-help videos and 'how-to' communities. Miles comments: 'One of the big successes during the pilot, which I was hoping for but it surprised me nonetheless, was the amount of people who started to self-help and to help others.'[10]

Measuring Success

The impact of the Flexible Working programme has been rigorously measured right from the start with user data that could be cut in a range of ways such as job grade, division or device used for each tool. The team recorded impressive levels of adoption for the pilot with the fastest ever adoption of Webex within an organization.[11] Although uptake for Quad was lower, it was in line with expectations and roughly equivalent to the 90–9–1 rule for social media adoption.[12]

A survey of participants at the end of the six-month pilot returned positive results in areas such as work–life balance, productivity, ease and effectiveness of homeworking, and quality of working relationships, as demonstrated by the following responses:[13]

- improved my work–life balance (53 per cent);

- made me more productive (77 per cent);

- helped me to strengthen existing relationships (67 per cent);

- enabled me to form new relationships (48 per cent);

- made homeworking easier and more effective (63 per cent);

- enabled me to work more effectively from locations other than my main office (65 per cent).

The team also reported a 6 per cent uplift in employee engagement for the pilot group, one of the highest increases ever recorded up to that point.

Notes

1   This case study is based on publically available sources – see references that follow.
2   Cisco (13 February 2012) Virgin Media enables flexible working with Cisco Quad collaboration
        software. Cisco, The Network press release: http://newsroom.cisco.com/press-release-
        content?articleId=666430 [accessed 07.05.14].
3   CSCOPR (10 February 2012) Virgin Media enables flexible working with Cisco Quad collaboration
        software. CSCOPR on YouTube: https://www.youtube.com/watch?v=8_jD957CWq0 [accessed
        07.05.14].
4   Bynghall, Steve (2013). Digital Workplace Fundamentals: The integrated approach. Digital Workplace
        Group: http://www.digitalworkplacegroup.com/resources/download-reports/digital-workplace-
        fundamentals [accessed 07.05.14].
5   Thomson Reuters IDS (June 2012) Virgin Media – flexible working 2012. IDS HR in Practice study:
        https://ids.thomsonreuters.com/hr-in-practice/case-studies/virgin-media---flexible-working-2012
        [accessed 07.05.14].
6   Cisco (2012) Virgin Media strengthens innovation by connecting employees using WebEx Social,
        WebEx Meetings, and TelePresence. Cisco Customer Case Study: http://www.cisco.com/cisco/web/UK/
        casestudies/assets/pdfs/virgin_media_case_study.pdf [accessed 07.05.14].
7   Peer Awards for Excellence (2012) Flexible working enabled by enterprise social media. Peer Award
        for Technology – Virgin Media in association with Cisco: http://thepeerawards.com/12-045-virgin
        [accessed 07.05.14].
8   As 5 above.
9   Stillman, Jessica (13 February 2013) The lessons of Virgin Media's flexible working initiative. Gigaom:
        http://gigaom.com/2012/02/13/the-lessons-of-virgin-medias-flexible-working-initiative [accessed
        07.05.2014].
10   As 9 above.
11   As 6 above.
12   Wikipedia (2014) 1% rule (Internet culture). http://en.wikipedia.org/wiki/1%25_rule_%28Internet_
        culture%29 [accessed 07.05.2014].
13   As 5 above.

Index

Bold page numbers indicate figures, *italic* numbers indicate tables.

A

absenteeism reductions 105–6
Accenture 30
accessibility
    of leadership 50–2
    of system 136
addiction
    digital 57–8
    work 56–7
Addison Lee 11
Adobe 17
adoption of new services 123
Agarwal, Anant 63–4
age as less relevant to work 43
Agile Working Programme, Unilever
    158–9, 162, 171
Alaska Airlines 81
Alexander, Christopher 13
Alpine Access 105
American Electric Power 112
American Express 178
Andraka, Jake 67–8
Angelou, Maya 39
application programming interfaces
    (APIs) 132
artisans 7–8
Asian Paints 151
AT&T 111
automation of work 40, 59–60
Automattic 36–7, 44
autonomy over own work 7, 8

B

banking via mobile phones 59
Barclays Bank 42–3
    MyZone 8–9
beauty in the digital world 17
benefits of digital workplaces. *see*
    business case for digital workplaces
Benioff, Marc 51–2
Bloom, Louise 123
Bougues Telecom 15–16
branding 133, 135
Branson, Richard 50
Brynjolfsson, Erik 39, 45, 58
BT 105, 107
budget
    for programmes 149–50
    strategy 154
BUPA 111
business case for digital workplaces
    absenteeism reductions 105–6
    consolidation of systems and processes
        100–1
    continuity of business 97, 107–9
    corporate social responsibility (CSR)
        98, 109–13
    cost optimization 97, 100–3
    environmental impact, reduction of
        109–10
    examples 100–1, 101–2, 102–3, 105,
        106, 107, 108–9, 110, 111, 112–13,
        183–9
    innovation acceleration 98, 111–13
    presenteeism reductions 106
    productivity of staff 97, 103–5

real estate reductions 101–2
revenue increases 98, 110–11
staff turnover reductions 106–7
strategic business alignment 98–9
summary of benefits 97–8
travel reductions 102–3
Virgin Media 204
business objectives
digital workplace, alignment with
98–9
strategy for a digital workplace 152–3
buy-in to programmes 149–50
Bynghall, Steve 142

C
Campbell, Danette 156, 162–3
capability, digital. *see* maturity, digital
Capgemini 110
Carlyle, Thomas 33
Cemex 112
change management 154–5, 161–3, 201–2
Chief Digital Officers (CDOs) 150–1
Chubb 81
Cisco 33–4, 102–3, 129, 157, 160, 177,
191–6
Citi 102
Clarkson, Martha 197–200
co-working spaces 28–9
Coco-Cola Enterprises 35
coherence 90, *91*
collaboration
Automattic as example 36–7
as benefit of no physical office 30
business objectives alignment with
digital workplace 98–9
as continuous 34
effectiveness of digital 33–4
extent of via technology 33
maturity, digital 87, *88*
Shell as example 35–6
spaces for 199

training in digital teamwork 34–5
trust 36
communication
levels of maturity 85, *86,* 87
strategy 155
community 87, *88*
compliance management 160–1
connection addiction 57–8
connection as benefit of no physical office
30
ConocoPhillips 176
consolidation of systems and processes
100–1
content 133, 135
continuity of business 97, 107–9
control over own work 7, 8
corporate social responsibility (CSR) 98,
109–13
cost optimization 97, 100–3
cross-functional delivery teams 145–9,
147, 148, 158–9, 203
culture, organizational 154
customers, impact of digital workplace
on 19
customization 132

D
data feeds and APIs 132
Day, Helen 142
debt, user experience 125
delivery teams, cross-functional 145–9,
147, 148, 158–9, 203
design of digital workplaces
branding 133
Cisco 129, 191–6
communication to stakeholders 129
content 133
data feeds and APIs 132
debt, user experience 125
designed/non-designed workplace
122–3

effects of poor user experience 123–5
enterprise search 132–133
failings in 121–2
focus on user experience 128–36
fragmentation as current experience
     120–1
iteration and intervention 131
mental models 128
non-designed digital workplace
     122–3, 131–2
participatory 129
personalization and customization
     132
personas 129–31
poor user experiences 117–18
procurement 133–6
and productivity 123–4
research, user experience 128
retention of staff 124
risky behaviours 124
satisfaction, employee 124
scenarios 129
solutions to user experience problems
     127–37
specifications, used-centred 133
stakeholder workshops 129–30
standards for digital workplaces
     125–7
strategy, user experience 128
structure and navigation 132
testing with users 131
and user experiences 119–20
vendor management 136–7
visual design 133
wireframe diagrams 131
design of offices and style of working 25
Deutsche Telekom 58
Digital Renaissance, The 3–5
     Protestant work ethic, end of 6–7
     work applicable to 10–11
digital work ethic 6–7, 8

*Digital Workplace - How technology is
     liberating work, The* (Miller) 78–9
Digital Workplace Group (DWG) 27–8,
     30
digital world
     beauty in 17
     compared to physical world 13–14
     unsatisfactory state of 14–15
Dow Chemical 50–1

E
education
     change, acceleration in needed 64
     equality, global 65
     Essa Academy 66–7
     family relationships, benefits to 69
     flexibility in 67–9
     Khan Academy 65, 66
     Massive Open Online Course (MOOC)
          platforms 65–6
     redesign of 65
     schools, evolution in 69–70
     self-organized learning environments
          (SOLE) 67–9
     skills gaps 63, 64
     teachers' role, evolution in 69
     Thiel Fellowship 68–9
Einstein, Albert 63
email, restriction of 58
employees
     disappointment in system 18–19
     productivity of 97, 103–5, 123–4, 198–9
     retention of 106–7, 124
     satisfaction of 124, 174
     turnover reductions 106–7
enterprise search 132–133
environmental impact, reduction of
     109–10
E.ON 110
Ernst & Young 110
Essa Academy 66–7

ethics, work 6–7, 8
expertise-finders 112

F
Fife Council 105
flexibility
    culture and policies to support 104
    maturity, digital 90, *92*, 93
fragmentation as current experience
        120–1, 142
Freed, Ephraim 79
freelancing 43–4
Fried, Jason 23, 29

G
Gino, Francesca 169
glasses
    Google 24, 81
    smart 104
GlaxoSmithKline 102
Goebel, Nancy 103
Google 24
    glasses 81
governance model for digital workplaces
        155–61
Greenfield, Dave 57–8

H
HCL Technologies 111
Heine Brothers coffee shops 16
Heinemeier Hansson, David 23, 29
Hewlett Packard 24
home offices, increase in 26

I
IBM 102, 177
ideas management systems 112
implementation of digital workplaces. *see*
        programmes, digital workplace
Industrial Revolution 9–10
information sharing 85, *86, 87*

innovation acceleration 98, 111–13, 198–9
insourcing of work 44
integrated approach to implementation
    budget for 149–50
    buy-in to programmes 149–50
    cross-functional delivery teams 145–9,
        147, 148
    fragmentation 142–3
    importance of 142–3
    leadership 150–1
    scope of programme 143
    sponsorship by senior management
        151
    stakeholder involvement 143–5, **145**
    teams, digital 151–2
    terminology 149–50
    visualization 150
Intel 160
intranets
    as core component 78
    and innovation 112
    metrics for measurement 173
    standards for 126
    as starting point 77
    strategy 153
    top-down 87
isolation in digital workplaces 55–6
ITAMCO 80
iteration of development 177–8

J
JetBlue Airways 17
jobs
    fixation on 41
    see also work
Jobs, Steve 17

K
Kennedy, Louise 121
key performance indicators (KPIs)
        168–71, *170*

Khan Academy 65, 66

L
Laird, Fiona 25
Lakha, Julie 149, 169
Lanier, Jaron 55, 120
Lao Tzu 49
leadership
    accessibility of 50–2
    challenges faced 49–50
    digital, examples of 50–2
    digital communication skills needed
        52–3
    lack of, digitally 50
    programmes, digital workplace 150–1
    Servant Leadership 53
    and trust 52–3
learning. *see* education
Leary, Timothy 29
levels of maturity
    communication and information 85,
        86, 87
    community and collaboration 87, *88*
    defined *85*
    mobility and flexibility 90, *92*, 93
    services and workflow *89*, 89–90
    structure and coherence 90, *91*
Lilly 112–13
LiveOps 111
Liveris, Andrew 50–1
Liverpool University 80
local governance 159–60
locations for meetings
    co-working spaces 28–9
    nearer to staff 27

M
machines, replacement of work by 40,
        59–60
Malmo 178
management, sponsorship by 151, 156

Massive Open Online Course (MOOC)
        platforms 65–6
maturity, digital
    aims of model 84
    communication and information 85,
        86, 87
    community and collaboration 87, *88*
    levels of maturity 84, *85*
    mobility and flexibility 90, *92*, 93
    readiness of digital workplaces 93–4
    services and workflow *89*, 89–90
    structure and coherence 90, *91*
Mayer, Marissa 24, 50
McAfee, Andrew 39, 45, 58
McQueen, Steve 3
measurement of progress and
        performance
    Cisco 195–6
    continuous 171–2
    as continuous 177–8
    evolving and embedding 178–80
    frameworks for 171–2
    importance of 167
    individual projects 172–6, *175*
    key performance indicators (KPIs)
        168–71, *170*
    lack of 167
    metrics for 169, *170*, 172–6, *175*
    success stories, use of for 176–7
    Unilever 171
    Virgin Media 205–6
*Measuring Intranets: A guide to intranet
        metrics and measurement* (Tubb)
        173
medical technologies 59
meetings
    co-working spaces 28–9
    increased significance of 55–6
    nearer to staff 27
    remote 102–3
mental models 128

metrics for measurement 169, *170, 172–6, 175*
Microsoft 18, 23, 110, 197–202
Miller, Paul 78–9
Mitra, Sugata 64, 67
Mitsubishi Electric 80
mobile phones, banking via 59
mobility
    and accessibility 136
    maturity, digital 90, *92,* 93
modes of work 25
Motley Fool, The 44
MyZone at Barclays Bank 8–9

N
NASA 102
National Institute of Health (NIH) 109
navigation 132, 136
Netherlands Defence Force 101
New Economics Foundation (NEF) 45
NHBC 106
non-designed digital workplaces 122–3, 131–2

O
office, physical. *see* physical world
older/younger workers 42–3
organizational culture 154
Ottawa 106
ownership by senior management 151, 156

P
participatory design of digital workplaces 129
performance checking. *see* measurement of progress and performance
personalization and customization 132
personas 129–30
personnel. *see* employees

physical world
    benefits of no physical office 30
    closure of office of DWG 27–8
    co-working spaces 28–9
    compared to digital world 13–14
    future of workplace in 23
    home offices, increase in 26
    meeting locations close to staff 27
    new style workplaces 29–30
    revenge of the office 24
    style of working and office design 25
Pichette, Patrick 24
Plummer, Jacobina 159
policies 160
power over own work 7, 8
presenteeism reductions 106
printing, invention of 3
PRISA 151, 179–80
processes, systems and, consolidation of 100–1
procurement
    branding and appearance 135
    checklist 134–6
    collaboration with specialists 133
    content compatibility 135
    mobility and accessibility 136
    navigation and integration 136
    specifications, used-centred 133
    standards for 133
    testing of off-the-shelf products 134
    user authentication and support 134–5
    and user experience 133–6
    vendor management 136–7
productivity of staff 97, 103–5, 123–4, 198–9
programmes, digital workplace
    budget for 149–50
    buy-in to 149–50
    change management 161–3
    checklist for 163–4
    continuous improvement 177–8

cross-functional delivery teams 145–9, **147, 148**
governance model 155–61
governance model for the digital workplace 155–61
implementation 161–2
integrated approach 142–52
leadership 150–1
roll-out 161
scope of 143
stakeholder involvement 143–5, **145**
strategy 152–5
teams, digital 151–2
terminology 149–50
visualization 150
see also measurement of progress and performance
progress checking. *see* measurement of progress and performance
Protestant work ethic, end of 6–7
PwC 35, 100, 105, 161, 174, 176, 183–9

Q
quantified self technologies 59

R
Race Against the Machine (Brynjolfsson and McAfee) 40
readiness of digital workplaces 93–4
real estate
    input from 144
    reductions in 101–2
redesign of work 41
Regus 28–9
RehabCare 107
Remote: Office not required (Fried and Heinemeier Hansson) 29
Renaissance, The 3, 9
research
    ongoing 177–8
    strategy 154

into user experience 128, 178
retention of staff 106–7, 124
revenge of the office 24
revenue increases 98, 110–11
risk management 160–1
Rosso, Sara 37
Ryan LLC 107

S
Salesforce.com 51–2
satisfaction, employee 124, 174
scenarios 129
schools. *see* education
scope of programme 143
Second Machine Age, The (Brynjolfsson and McAfee) 39, 58
self-driving cars 59–60
self-organized learning environments (SOLE) 67–9
senior management, sponsorship by 151, 156
Servant Leadership 53
services and workflow, maturity of *89,* 89–90
setting up digital workplaces. *see* programmes, digital workplace
Shell 35–6, 101
Sidetracked (Gino) 169
skills gaps 63, 64
Spark 176, 183–9
specifications, used-centred 133
sponsorship by senior management 151, 156
staff. *see* employees
stakeholders
    and design of digital workplaces 129
    involvement of 143–5, **145,** 157–8
standards
    for digital workplaces 125–7
    governance 160
    procurement 133

steering groups 157, 159
Stora Enso 147, 149
strategy
    budget 154
    business, alignment with digital
        workplaces 98–9
    business needs 154
    change management 154–5
    communication 155
    for a digital workplace 152–5
    intranets 153
    organizational culture 154
    planning 155
    research 154
    suppliers 155
    user experience 128
structure and coherence, maturity of 90, 91
structure and navigation 132
style of working 25, 191–2
success stories 176–7
suppliers 155
systems and processes, consolidation of
    100–1

T
teachers' role, evolution in 69
teams
    cross-functional 145–9, **147, 148,**
        158–9, 203
    digital 151–2
teamwork, digital
    Automattic as example 36–7
    Shell as example 35–6
    training and guidance in 34–5
    trust 36
teleconferencing
    solutions 17–18
    travel reductions via 102–3
testing systems with users 131
Thiel Fellowship 68–9
third places 28–9

Tony Roma restaurant chain 81
traditional companies 25
training 34–5, 160, 163, 174
travel, reductions in 102–3
Tredway Lunsdaine and Doyle LLP 81
trust
    in digital workplaces 36
    and leadership 52–3
Tubb, Chris 83, 129, 151
turnover, staff, reductions in 106–7

U
Unilever 23, 25, 80, 142, 146, 148, 158–9,
    162, 171
United States Air Force Central
    Adjudication Facility 105
United States Defense Information
    Systems Agency (DISA) 108–9
United States government 109
United States Marine Corps 109
United States Patent and Trademark
    Office (USPTO) 25–6, 81, 101, 108,
    142, 147, 149, 156–7, 162–3
user authentication and support 134–5
user experiences
    adoption of new services 123
    branding 133
    Cisco 129, 191–6
    content 133
    control over, level of 122–3
    data feeds and APIs 132
    debt, user experience 125
    definition of good 119–20
    and design of digital workplaces
        119–20
    designed/non-designed workplaces
        122–3
    effects of poor 123–5
    enterprise search 132
    focus on in design of digital
        workplaces 128–36

mental models 128
non-designed digital workplaces
    131–2
participatory design of digital
    workplaces 129
personalization and customization
    132
personas 129–30
poor 117–18
procurement 133–6
and productivity 123–4
research 128, 129, 178
retention of staff 124
risky behaviours 124
satisfaction, employee 124, 174
scenarios 129
solutions to problems with 127–37
specifications, used-centred 133
stakeholder workshops 129
strategy 128
structure and navigation 132
testing systems with users 131
Virgin Media 204–5
visual design 133
wireframe diagrams 131

V
vendor management 136–7
video conferencing
    solutions 17–18
    travel reductions via 102–3
Virgin Media 34–5, 106, 161, 203–6
virtual meetings 17–18, 102–3
visual design of systems 133
visualization 150
Vodafone UK 103
Volkswagon 58

W
Walmart 177
Warrior, Padmasree 14–15

wearable technology 104
Wentworth, Dianne 43
Whitman, Meg 24
Wiltshire Council 147, 149
wireframe diagrams 131
work
    adaptations, need for 60
    addiction to 56–7
    allocation of 45
    amount of per person 44–6
    automation of 40
    Digital Renaissance examples 10–11
    email, restriction of 58
    freelancing 43–4
    impact of Digital Renaissance 5–6
    insourcing of 44
    no shortage of 41–2
    Protestant work ethic, end of 6–7
    redesign of 41
    restructuring of 43–4
    revolutions in 39–40
    stretching of age range 42–3
    style of working 25, 191–2
    working week, redesign of 45–6
    younger/older workers 42–3
work ethics 6–7, 8
workflow, maturity of *89*, 89–90
working groups 157–8, 159
working week, redesign of 45–6
Workplace Advantage programme,
    Microsoft 197–202
workplaces, digital
    addiction to work 56–7
    attention being paid to 81–2
    Bougues Telecom 15–16
    customers, impact on 19
    definition 78–80
    development of services for 17
    drivers of 82
    email, restriction of 58
    examples 80–1

failings in 121–2
fragmentation as current experience
        120–1
governance model for 155–61
Heine Brothers coffee shops 16
impact on work and life 79–80
integrated approach, need for 82–3
intranets 77, 78
isolation in 55–6
motivation for 16
promise of 80–1
readiness of 93–4
staff disappointment in system 18–19
standards for 125–7
telepresence solutions 17–18
trust 36
unsatisfactory current state of 15

vision and standards, need for
        19–20
see also business case for digital
        workplace; measurement of
        progress and performance;
        programmes, digital workplace

X
Xerox 131

Y
Yahoo! 24
Young, Paula 176, 183–9
younger/older workers 42–3

Z
Zara 10–11